Probability and Mathematical Statistics (Continued)

PRESS • Bayesian Statistics
PURI and SEN • Nonparametric Methods in General Linear Models
PURI and SEN • Nonparametric Methods in Multivariate Analysis
PURI, VILAPLANA, and WERTZ • New Perspectives in Theoretical and Applied Statistics
RANDLES and WOLFE • Introduction to the Theory of Nonparametric Statistics
RAO • Linear Statistical Inference and Its Applications, *Second Edition*
RAO • Real and Stochastic Analysis
RAO and SEDRANSK • W.G. Cochran's Impact on Statistics
RAO • Asymptotic Theory of Statistical Inference
ROBERTSON, WRIGHT and DYKSTRA • Order Restricted Statistical Inference
ROGERS and WILLIAMS • Diffusions, Markov Processes, and Martingales,Volume II: Îto Calculus
ROHATGI • An Introduction to Probability Theory and Mathematical Statistics
ROHATGI • Statistical Inference
ROSS • Stochastic Processes
RUBINSTEIN • Simulation and The Monte Carlo Method
RUZSA and SZEKELY • Algebraic Probability Theory
SCHEFFE • The Analysis of Variance
SEBER • Linear Regression Analysis
SEBER • Multivariate Observations
SEBER and WILD • Nonlinear Regression
SEN • Sequential Nonparametrics: Invariance Principles and Statistical Inference
SERFLING • Approximation Theorems of Mathematical Statistics
SHORACK and WELLNER • Empirical Processes with Applications to Statistics
STOYANOV • Counterexamples in Probability

Applied Probability and Statistics

ABRAHAM and LEDOLTER • Statistical Methods for Forecasting
AGRESTI • Analysis of Ordinal Categorical Data
AICKIN • Linear Statistical Analysis of Discrete Data
ANDERSON and LOYNES • The Teaching of Practical Statistics
ANDERSON, AUQUIER, HAUCK, OAKES, VANDAELE, and WEISBERG • Statistical Methods for Comparative Studies
ARTHANARI and DODGE • Mathematical Programming in Statistics
ASMUSSEN • Applied Probability and Queues
BAILEY • The Elements of Stochastic Processes with Applications to the Natural Sciences
BARNETT • Interpreting Multivariate Data
BARNETT and LEWIS • Outliers in Statistical Data, *Second Edition*
BARTHOLOMEW • Stochastic Models for Social Processes, *Third Edition*
BARTHOLOMEW and FORBES • Statistical Techniques for Manpower Planning
BATES and WATTS • Nonlinear Regression Analysis and Its Applications
BECK and ARNOLD • Parameter Estimation in Engineering and Science
BELSLEY, KUH, and WELSCH • Regression Diagnostics: Identifying Influential Data and Sources of Collinearity
BHAT • Elements of Applied Stochastic Processes, *Second Edition*
BLOOMFIELD • Fourier Analysis of Time Series: An Introduction
BOX • R. A. Fisher, The Life of a Scientist
BOX and DRAPER • Empirical Model-Building and Response Surfaces
BOX and DRAPER • Evolutionary Operation: A Statistical Method for Process Improvement
BOX, HUNTER, and HUNTER • Statistics for Experimenters: An Introduction to Design, Data Analysis, and Model Building

Applied Probability and Statistics (Continued)

BROWN and HOLLANDER • Statistics: A Biomedical Introduction
BUNKE and BUNKE • Statistical Inference in Linear Models, Volume I
CHAMBERS • Computational Methods for Data Analysis
CHATTERJEE and HADI • Sensitivity Analysis in Linear Regression
CHATTERJEE and PRICE • Regression Analysis by Example
CHOW • Econometric Analysis by Control Methods
CLARKE and DISNEY • Probability and Random Processes: A First Course with Applications, *Second Edition*
COCHRAN • Sampling Techniques, *Third Edition*
COCHRAN and COX • Experimental Designs, *Second Edition*
CONOVER • Practical Nonparametric Statistics, *Second Edition*
CONOVER and IMAN • Introduction to Modern Business Statistics
CORNELL • Experiments with Mixtures: Designs, Models and The Analysis of Mixture Data
COX • Planning of Experiments
COX • A Handbook of Introductory Statistical Methods
DANIEL • Biostatistics: A Foundation for Analysis in the Health Sciences, *Fourth Edition*
DANIEL • Applications of Statistics to Industrial Experimentation
DANIEL and WOOD • Fitting Equations to Data: Computer Analysis of Multifactor Data, *Second Edition*
DAVID • Order Statistics, *Second Edition*
DAVISON • Multidimensional Scaling
DEGROOT, FIENBERG and KADANE • Statistics and the Law
DEMING • Sample Design in Business Research
DILLON and GOLDSTEIN • Multivariate Analysis: Methods and Applications
DODGE • Analysis of Experiments with Missing Data
DODGE and ROMIG • Sampling Inspection Tables, *Second Edition*
DOWDY and WEARDEN • Statistics for Research
DRAPER and SMITH • Applied Regression Analysis, *Second Edition*
DUNN • Basic Statistics: A Primer for the Biomedical Sciences, *Second Edition*
DUNN and CLARK • Applied Statistics: Analysis of Variance and Regression, *Second Edition*
ELANDT-JOHNSON and JOHNSON • Survival Models and Data Analysis
FLEISS • Statistical Methods for Rates and Proportions, *Second Edition*
FLEISS • The Design and Analysis of Clinical Experiments
FLURY • Common Principal Components and Related Multivariate Models
FOX • Linear Statistical Models and Related Methods
FRANKEN, KÖNIG, ARNDT, and SCHMIDT • Queues and Point Processes
GALLANT • Nonlinear Statistical Models
GIBBONS, OLKIN, and SOBEL • Selecting and Ordering Populations: A New Statistical Methodology
GNANADESIKAN • Methods for Statistical Data Analysis of Multivariate Observations
GREENBERG and WEBSTER • Advanced Econometrics: A Bridge to the Literature
GROSS and HARRIS • Fundamentals of Queueing Theory, *Second Edition*
GROVES, BIEMER, LYBERG, MASSEY, NICHOLLS, and WAKSBERG • Telephone Survey Methodology
GUPTA and PANCHAPAKESAN • Multiple Decision Procedures: Theory and Methodology of Selecting and Ranking Populations
GUTTMAN, WILKS, and HUNTER • Introductory Engineering Statistics, *Third Edition*
HAHN and SHAPIRO • Statistical Models in Engineering
HALD • Statistical Tables and Formulas
HALD • Statistical Theory with Engineering Applications

(*continued on back*)

Most texts on model building present tidy methodological packages of information which have limited application for dealing with problems as they are encountered in day-to-day activities. Consequently, scientific and business professionals are confronted with the perplexing dilemma of trying to solve real-world problems using straight textbook-taught solution formats which often simply do not work.

Emphasizing "a problem in search of a solution" rather than a "methodology in search of an application" approach to the construction of practical models, *Empirical Model Building* approaches the formulation of real-world problems and gives insight into the construction of desktop computer models in a wide variety of situations. Using more than twenty case studies, the book presents concepts designed to help professionals formulate and solve problems in actual day-to-day situations.

Chapter 1 details models of growth and decay, starting with an examination of the "privatized" social security system advocated by some people. It shows how the advent of the calculator and the computer should cause us to consider alternatives to the "closed form solution." The book also covers the motivations for the tax reforms of 1981, the spread between the actual value and the nominal value of a mortgage, population growth, and the consequences of mutation and natural selection in the development of resistance to chemotherapeutic agents.

Chapter 2 looks at more complicated systems in which competition and interaction add to the complexity of the model. Chapter 3 argues that the introduction of the microprocessor should change fundamentally the way we carry out modelling. Here, simulation is advocated as an aggregation alternative to the "closed form."

Chapter 4 addresses some nonclassical methods of data analysis which are highly interactive with the human visual perception system. Finally, Chapter 5 takes a contrary approach to the way several well-established theories are perceived. The primary point here is that we tend to pay too much attention to the mathematical consequences of axiomitized descriptions of real-world systems without checking carefully the adequacy of the axioms used to describe these systems.

By exposing students of statistics as well as scientific and business professionals to a variety of nonstandard modelling situations, *Empirical Model Building* utilizes the synthesizing powers of the mind to enable readers to apply what they have learned to other unrelated problems as they occur in the work environment.

About the author

James R. Thompson is Professor and Chairman of the Department of Statistics at Rice University and Adjunct Professor of Biomathematics at the University of Texas M. D. Anderson Cancer Center. He is a Fellow of the American Statistical Association, a Fellow of the Institute of Mathematical Statistics and an elected Member of the International Statistical Institute. Dr. Thompson is the co-author of *Nonparametric Probability Density Estimation* and the co-editor of *Cancer Modelling.* He has authored or co-authored some sixty articles on various aspects of modelling and statistics. He earned his BE degree in chemical engineering at Vanderbilt University and his PhD in mathematics at Princeton University.

Empirical Model Building

JAMES R. THOMPSON

Professor of Statistics
Rice University
Adjunct Professor of Biomathematics
University of Texas
M. D. Anderson Cancer Center

JOHN WILEY & SONS

New York • Chichester • Brisbane • Toronto • Singapore

Library of Congress Cataloging in Publication Data:

Thompson, James R. (James Robert), 1938–
Empirical model building/James R. Thompson.
p. cm.—(Wiley series in probability and mathematical statistics. Probability and mathematical statistics, ISSN 0271–6232)
Bibliography: p.
ISBN 0-471-60105-5
1. Experimental design. 2. Mathematical models. 3. Mathematical statistics. I. Title. II. Series.
QA279.T49 1989
519.5—dc19 88-20549
CIP

Printed in the United States of America

10 9 8 7 6 5 4 3 2 1

To
My Mother, Mary Haskins Thompson

Preface

Empirical model building refers to a mindset that lends itself to constructing practical models useful in describing and coping with real-world situations. It does not refer to quick and dirty methods, which are used simply because they are ones we understand and we have them readily available in the form of off-the-shelf software. The fraction of real situations which can appropriately be addressed using, say, a linear regression package or a Newton's optimization routine or an autoregressive moving average forecasting procedure is small indeed. Most successful consultants are aware of this fact, but it escapes the attention of most academics and the orientation of most textbooks.

It is a recurring experience of students who receive their B.A. or their M.S. or their Ph.D. and journey forth into the real world of science, commerce, and industry, that they seldom find any problem quite like those they have been trained to solve in the university. They find that they have been studying tidy methodological packets of information which have rather limited application. If they are fortunate, they will accept the situation and begin to learn model building "on the job" as it were. On the other hand, it does appear a pity that so important a subject be relegated to the school of hard knocks. Consequently, a number of universities have experimented for the last 20 years or so with courses in model building.

The problem with most such courses and with books used in the courses is that they tend to focus on methodologies well understood by the instructor. After all, operations research, numerical analysis, statistics, and physical mathematics are all supposed to be branches of applied mathematics, and so anything presented from such a field is deemed applied. There are model building books that emphasize linear programming, others that emphasize queueing theory, and so on. The fact is, of course, that most "applied mathematics" is as divorced from the real world as algebraic topology and ring theory. Methodology, perceived as an end in itself, is usually a closed system that looks inward rather than a means to some end in the world outside the methodology.

Consultation is a subject very much related to model building. Increasingly, courses on consultation are included in the graduate curricula of departments of statistics. Short courses are frequently given on the topic to practicing professionals. The point of view of some of the instructors of such courses and the authors of texts for such courses seems to be that bedside manner and psychological support are the main contributions of a consultant. When listening to some of the case studies and the recommended approach of the instructors who use them, I have been struck by the similarities with recommended care for the terminally ill. Most consulting clients for whom I have worked over the years do not want tea and sympathy. They want results. They have a problem (or problems) which they usually cannot quite articulate. They want the consultant to formulate the problem and solve it—nothing more nor less. They are completely unmoved by statements on the part of the consultant that "that is not my field." Most real-world consulting jobs are not anybody's field. They are problems that have to be attacked *de novo*. The ability to handle consulting problems is deemed by some to be "a gift." Some people are supposed to have the facility, whereas others do not, though they may be expert theoreticians. My experience is that model building is an attractive means whereby one develops consulting skills. I have former students who call and write about this or that off-the-wall problem which they were given and were able to formulate and solve thanks to the insights they gained in the Rice University model building course.

The purpose of the course in model building, which I have taught at Rice University for 10 years, has been to try to start with problems rather than with methodologies. The fact that I am a statistician (and an undergraduate chemical engineer) certainly biases my approaches and to some extent the problems I choose to examine. However, the experience of a decade is that the approach is successful to a very considerable degree in preparing individuals to formulate and solve problems in the real world. The students in the course have ranged from Ph.D. candidates to sophomores.

We live in a world in which the collection of subjects required for this or that major may be only marginally relevant to the real world. This or that "hard-nosed" petroleum engineering curriculum, for example, may become as irrelevant to the job market as a curriculum in Etruscan archaeology. Of course, the advocates of a liberal arts education have always cheerfully accepted this fact and realistically informed students that such an education was not supposed to be job oriented. The argument that a diverse educational experience is better preparation than training in a narrow area which may become obsolete in a few years has some appeal. But between the extreme of a completely nontechnical education and one of narrow specialization lies a considerable middle ground. Students have begun to understand that obsolescence can overtake any field. Many expect that they will experience several career changes during their professional lives. Accordingly, model building can be looked on as a kind of "liberal sciences" subject. By exposing the student to a variety of modeling situations, it is to be hoped that the synthesizing powers of the human brain will prepare him or her for other situations that are covered neither by other courses nor by the model building course. Experience indicates that this indeed is the case.

In a classroom setting, I form students into two- and three-person groups, which address 10 model building scenarios, each related to one of the sections in the book. The use of group reports as the primary vehicle of evaluation is another attempt to make the course more representative of situations faced in the real world. A week or two is allowed for the preparation of each report. Although the course is one semester in duration at Rice University, the fact that there are more than twice as many sections in the book as can be covered in a semester provides for selections of material based on the preferences of the instructor and the particular composition of the group.

Chapter 1 is concerned with several models of growth and decay. Section 1.1 is an attempt to look at the kind of "privatized" social security system advocated by some. It shows, among other things, how the advent of the calculator and the computer should cause us to consider alternatives to the "closed form solution." In this section, some advantages of the "do loop" are given. Section 1.2 examines the motivation for the tax reforms of 1981. Section 1.3 examines the spread between the actual value and the nominal value of a mortgage. Section 1.4 departs from rather well-defined accounting models and goes into the more ambiguous country of population growth. Section 1.5 discusses the consequences of metastatic progression and the development of resistance to chemotherapeutic agents in the treatment of cancer.

Chapter 2 looks at rather more complicated systems in which competition and interaction add to the complexity of the model. Section 2.1 is a highly speculative analysis of the population of ancient Israel. Part of the motivation for this section is to demonstrate how a supposedly sterile data set, like that in the Book of Numbers, may become quite significant when analyzed in the light of a model. Section 2.2 shows the enormous advance in data analysis caused by the data compression technique of John Graunt. Section 2.3 examines some considerations in the modeling of combat situations. More particularly, a model-based argument is made as to the possible value of fortifications in modern combat (General Patton notwithstanding). Section 2.4 shows how the predator–prey model first advocated by Volterra can not only be used to model competition of species but can also be applied to the dramatically different situation of the body's immune response to cancer.

Section 2.5 considers the relatively trivial subject of pyramid clubs as a precursive analysis of epidemics. Section 2.6 examines the AIDS epidemic and gives an analysis that suggests that the epidemic might have been avoided if public health authorities had simply closed down establishments, such as bathhouses, which encourage high contact rate homosexual activity.

Chapter 3 argues further that the advent of the microprocessor should change the way we carry out modeling. Here, simulation is advocated as an aggregation alternative to the closed form. Section 3.1 examines the use of simulation as an alternative to numerical approximations to the closed form for the point-wise evaluation of models generally described by differential equations. This concept was Johann von Neumann's original motivation for construction of the digital computer, but the point is made in Section 3.1 that computing has reached the speed and cheapness where we should, in many cases, dispense with the differential equation formulation altogether and go rather from the axioms to the pointwise evaluation of the function. Section 3.2 shows a computer intensive, but conceptually simple, procedure whereby we can use a data set to construct many "quasi data" sets. Section 3.3 considers the use of simulation-based alternatives for the estimation of parameters characterizing stochastic processes. The SIMEST algorithm dealt with in this section has made possible the modeling of processes not tractable using classical closed form techniques.

Chapter 4 addresses some nonclassical methods of data analysis which are highly interactive with the human visual perception system. Both fall, more or less, under the Radical Pragmatist approach mentioned in the Introduction. Section 4.1 attempts to give an analytical description of John W. Tukey's exploratory data analysis. Section 4.2 examines the challenges and problems associated with the analysis of higher-dimensional data via nonparametric density estimation.

Chapter 5 takes a contrarian approach to the way several well-established theories are perceived. The primary point made here is that we tend to pay too much attention to the mathematical consequences of axiomitized descriptions of real-world systems without checking carefully as to the adequacy of the axioms to describe these systems. Section 5.1 examines several problems concerned with group consensus, in particular the famous impossibility theorem of Kenneth Arrow. Section 5.2 examines Charles Stein's proof that the sample mean can always be improved on as an estimate for the mean of a normal distribution for dimensionality greater than two. Section 5.3 examines the fuzzy set theory of Lofti Zadeh. Section 5.4 argues that quality control, while a subject of great importance, is generally misunderstood, due to an improper anthropomorphization of machines and systems.

Appendix A.1 gives a brief introduction to stochastics. Appendix A.2 presents the robust optimization algorithm of Nelder and Mead.

In general, the material in the first two chapters is rather simpler than that in subsequent chapters. However, even the simplest sections ought not be despised, as they are generally nontrivial, even though the mathematical tools required for the formulation of the models and their analysis may be. Some of the sections state several results without proof, since to go into such areas as decision theory in detail would require an unnecessary and lengthy foray into complex, frequently marginal, areas of applied mathematical theory. Again, this practice reflects reality, for most of us, when attacking a new problem, will use some methodological results without starting at the beginning to prove them.

A subset of the book was presented in the annual 12 hour short course at the 1986 Army Design of Experiments Conference to a group of some 40 scientific professionals, where a preliminary version of the book was used in lecture note format. Several of these, and others who have read the manuscript subsequently, have been kind enough to read and comment on it. A number of their criticisms have been incorporated into the book, as have the many helpful comments and suggestions of the referee. I am grateful to Dr. Robert Launer and the Army Research Office (Durham) for the invitation to give the lectures and wish to acknowledge support by both the ARO and the ONR under DAAG-29-85-K-0212, DAAL-03-88-K0131 and N00014-85-K-0100, respectively.

The author wishes to acknowledge the encouragement and advice of Neely Atkinson, Robert Bartoszyński, Barry Brown, Diane Brown, Steve Boswell, Jagdish Chandra, Tim Dunne, Ed Johnson, Frank Jones, Tom Kauffman, Jacek Koronacki, Robert Launer, Ray McBride, Emanuel Parzen, Paul Pfeiffer, David Scott, Bea Shube, Malcolm Taylor, George Terrell, Patrick Tibbits, and John Tukey.

J. R. THOMPSON

Houston, Texas
September, 1988

Contents

Introduction

The study of mathematical models is closely connected to notions of scientific creativity. As of the present, there is no axiomatic or even well-defined discipline that is directly concerned with creativity. Even though we cannot display a progression of exercises that have as their direct objective the building of creativity, we can attempt to accomplish this goal indirectly. A mastery of a portion of Euclid's treatises on geometry does not appear directly to build up a potential statesman's ability to practice statecraft. Yet many effective statesmen have claimed that their studies of Euclid's geometry had helped achieve this effect. More directly, it is clear that the study of physics would likely be helpful in developing the ability to design good automobiles. It is this carry-over effect from one well-defined discipline to another less well-defined one which has traditionally been the background of scientific and technological education.

Valuable though an indirect approach to the gaining of creativity in a particular area may be, it carries with it certain dangers. We are rather in the same situation as the little boy who searched for his quarter, lost in a dark alley, under a bright streetlight on a main street. There is no doubt that the main street searching could be of great utility in the ultimate quest of finding the quarter. Many of the relevant techniques in quarter finding are similar, whether one is looking in the light or in the dark. Hopefully, the study of technique, albeit undertaken in a setting substantially different from that of the real problem, will be at least marginally useful in solving the real problem.

However, there is a natural temptation never to leave the comfort of idealized technique under the bright lights, never to venture into the murky depths of the alley where the real problem lies. How much easier to stay on the main street, to write a treatise on the topology of street lamps, gradually to forget about the lost quarter altogether.

In its most applied aspect, technique becomes problem solving. For example, if the little boy really develops a procedure for finding his particular quarter in the particular dark alley where he lost it, he will have been engaged in problem solving. Although it is difficult to say where problem posing ends and problem solving begins, since in the ideal state there is continuous interaction between the two, model building is more concerned with the former than with the latter. Whereas problem solving can generally be approached by more or less well-defined techniques, there is seldom such order in the problem posing mode. In the quarter finding example, problem posing would involve determining that it was important that the quarter be found and a description of the relevant factors concerning this task. Here, the problem posing is heuristic, difficult to put into symbols, and trivial. In the real world of science and technology, problem posing is seldom trivial but remains generally heuristic and difficult to put into symbols. For example, Newton's second principle states that force is equal to the rate of change of momentum or

$$F = \frac{d}{dt}(mv). \tag{0.1}$$

The solving of (0.1) for a variety of scenarios is something well defined and easily taught to a high school student. But the thought process by which Newton conjectured (0.1) is far more complex.

We have no philosopher's stone to unlock for us the thought processes of the creative giants of science. And we shall not explore the study of scientific biography much in this book. However, the case study approach appears to be useful in the development of creativity. By processes that we do not understand, the mind is able to synthesize the ability to deal with situations apparently unrelated to any of the case studies considered. It is the case study approach, historically motivated on occasion, which we shall emphasize.

At this point, it is appropriate that some attempt be made to indicate what the author means by the phrase *empirical model building.* To do so, it is necessary that we give some thought to the various ways in which scientists approach the concept of models. We shall list here only those three schools that appear to have the greatest numbers of adherents. The first group we shall term the *Idealists.* The Idealists are not really data oriented. They are rather concerned with theory as a mental process that takes a cavalier attitude toward the "real world." Their attitude can be summed up as follows: "If facts do not conform to theory, then so much the worse for the facts." For them, the "model" is all. An example of a pure Idealist is given by the character of Marat in Weiss' play *Marat-Sade.* Marat says, "Against Nature's silence I use action. In the vast indifference I invent a meaning." Idealists naturally have a hard time in the sciences. Sooner or later, the theories of a Lysenko, say, are brought up against the discipline imposed by reality and their work must surrender in the face of conflicting evidence. But "sooner or later" may mean decades. Once a theory has developed a constituency of individuals who have a vested interest in its perpetuation, particularly if the theory has no immediate practical implications, there will be a tendency of other scientists, who have no interest in the theory either way, to let well enough alone. No age, including this one, has been free of powerful Idealist enclaves.

The second group, that of the *Radical Pragmatists* (Occamites, Nominalists), would appear to be at the opposite end of the spectrum from that of the Idealists. The Radical Pramatists hold that data are everything. Every situation is to be treated more or less *sui generis.* There is no "truth." All models are false. Instead of model building, the Radical Pragmatist curve fits. He does not look on the fitted curve as something of general applicability, but rather as an empirical device for coping with a particular situation. The maxim of William of Occam was, *"Essentia non sunt multiplicanta praeter necessitatem";* roughly "The hypotheses ought not to be more than is necessary." The question here is what we mean by "necessary." All too frequently, it can happen that "necessary" means what we need to muddle through rather than what we need to understand. But few Radical Pragmatists would take the pure position of Weiss' de Sade, who says, "No sooner have I discovered something than I begin to doubt it and I have to destroy it again ... the only truths we can point to are the ever-changing truths of our own experience."

The *Realists* (Aristoteleans, Logocentricists, Thomists) might appear to some to occupy a ground intermediate to that of the Idealists and that of the Radical Pragmatists. They hold that the universe is governed by rational and consistent laws. Models, to the Realist, are approximations to bits and pieces of these laws. To the Realist, "We see through a glass darkly," but there is reality on the other side of the glass. The Realist knows his model is not quite right but hopes it is incomplete rather than false. The collection of data is useful in testing a model and enabling the Realist to modify it in appropriate fashion. It is this truth seeking, interactive procedure between mind and data which I term *empirical model building.*

To return again to Newton's second principle, the position of the Idealist might be that the old Newtonian formula

$$F = ma \tag{0.2}$$

is true because of an argument based on assumptions he finds compelling. But then we have the empirically demonstrable discovery of Einstein that mass is not constant but depends on velocity via

$$m = \frac{m_0}{\sqrt{1 - v^2/c^2}}. \tag{0.3}$$

The Idealist would have a problem. He might simply stick with (0.2). Or he could experience an intellectual conversion, saying, "Right, Einstein is correct; Newton is wrong. I am no longer a Newtonian but an Einsteinian" (or some less self-effacing dialectical version of the above conversion).

The reaction of the Radical Pragmatist might be, "You see, even an apparently well established model like Newton's is false. No doubt we will soon learn that Einstein's is false also. Both these models are useful in many applications, but their utility lies solely in their applicability in each situation."

The Realist is also unsurprised that Newton's model falls short of the mark. He notes that the discovery of Einstein will require a modification of (0.2). He readily accomplishes this by combining (0.1) and (0.3) to give

$$F = \frac{d}{dt}\frac{m_0 v}{\sqrt{1 - v^2/c^2}}. \tag{0.4}$$

The Realist views (0.4) as a better approximation to truth than (0.2) and expects to hear of still better approximations in the future.

The preceding should give the reader some feel as to what the author means by empirical model building (and also as to his prejudices in favor of the Realist position). It is the process that is sometimes loosely referred to as the "scientific method." As such, it has been around for millennia—although only for the last 500 years or so has quantitative data collecting enabled its ready use on nontrivial scientific problems. Realists might argue (as I do) that empirical model building is a natural activity of the human mind. It is the interactive procedure by which human beings proceed to understand portions of the real world by proposing theoretical mechanisms, testing these against observation, and revising theory when it does not conform to data. In any given situation, a scientist's empirical model is simply his current best guess as to the underlying mechanism at hand.

The Radical Pragmatist position has great appeal for many, particularly in the United States. There would appear to be many advantages to an orientation that allowed one to change his ground any time it was convenient to do so. But the ultimately nihilistic position of Radical Pragmatism has many practical difficulties. For example, data are generally collected in the light of some model. Moreover, from the standpoint of compression of information, a point of view which rejects truth also rejects uniqueness, causing no little chaos in representation. Finally, the old adage that, "He who believes in nothing will believe anything" appears to hold. The Radical Pragmatist seems to join hands with the Idealist more often than either cares to admit. There are certain groups who seem to wear the colors of both the Idealist and Radical Pragmatist schools.

The above taxonomy of contemporary scientists into three fairly well-defined schools of thought is obviously an oversimplification. Most scientists will tend to embody elements of all three schools in their makeup. For example, I might be (and have been) accosted in my office by someone who wishes me to examine his plans for a perpetual motion machine or his discovery of a conspiracy of Freemasons to take over the world. As a purely practical matter, because my time is limited, I will be likely to dismiss their theories as patently absurd. In so doing, I am apparently taking an Idealist position, for indeed I know little about perpetual motion machines or Freemasonry. But without such practical use of prejudice, nothing could ever be accomplished. We would spend our lives "starting from zero" and continually reinventing the wheel. There is a vast body of information which I have not investigated and yet take to be true, without ever carefully checking it out. This is not really "idealism"; this is coping. But if I read in the paper that Professor Strepticollicus had indeed demonstrated a working model of a perpetual motion machine, or if I heard that a secret meeting room, covered with Masonic symbols, had been discovered in the Capitol, then I should be willing to reopen this portion of my "information bank" for possible modification.

For similar practical reasons, I must act like a Radical Pragmatist more often than I might wish. If I see a 10 ton truck bearing down on me, I will instinctively try to get away without carefully investigating considerations of momentum and the likely destruction to human tissue as a result of the dissipation thereof. But I have the hope that the manufacturer of the truck has logically and with the best Newtonian theory in tandem with empirical evidence designed the vehicle and not simply thrown components together, hoping to muddle through.

In sum, most of us, while accepting the practical necessity of frequently assuming theories which we have not analyzed and using a great deal of instinctive rather than logical tools in our work, would claim to believe in objective reality and a system of natural laws which we are in a continual and evolutionary process of perceiving. Thus, most of us would consider ourselves to be Realists though we might, from time to time, act otherwise. Perhaps the minimal Realist maxim is that of Orwell's Winston Smith. "Freedom is the freedom to say that two plus two make four. If that is granted, all else follows."

CHAPTER ONE

Models of Growth and Decay

1.1. A SIMPLE PENSION AND ANNUITY PLAN

One easy introduction to the subject of growth models is obtained by considering an accounting situation, since such systems are relatively well defined. A pension and annuity plan is generally axiomitized clearly. To set up the model itself is consequently almost totally an exercise in problem solving.

Suppose that we set up a pension plan for an individual who starts working for a firm at age N_1, retiring at age N_2. The starting salary is S_1 at age N_1 and will increase at rate α annually. The employer and employee both contribute a fraction β of the salary each year to the employee's pension fund. The fund is invested at a fixed annual rate of return γ. We wish to find the value of the employee's pension fund at retirement. Furthermore, we wish to know the size of the employee's pension checks if he invests his pension capital at retirement in a life annuity (i.e., one that pays only during the employee's lifetime, with no benefit to his heirs at time of death). Let the expected death age given survival until N_2 be denoted by N_3.

Many investigators find it convenient to consider first a pilot study with concrete values instead of algebraic symbols. For example, we might try $S_1 = 20{,}000$; $N_1 = 21$; $N_2 = 65$; $\alpha = 0.02$; $\beta = 0.0705$; $\gamma = 0.05$; $N_3 = 75$. The values of α and γ are rather low, assuming a healthy, low inflation economy. The value of β is roughly the value used at present in the U.S. Social Security System. The value of N_2 is the same as the present regular Social Security retirement age of 65. No allowance is made for annual losses due to taxation, since pension plans in the United States generally leave the deposited capital untaxed until the employee begins his or her annuity payments (although the employee contributions to Social Security are taxed at full rate).

First, we note that at the end of the first year the employee will have approximately

$$P(1) = 2\beta S(1) = 2(0.0705)20{,}000 = 2820 \tag{1.1.1}$$

invested in the pension plan. This will only be an approximation, since most salaried employees have their pension fund increments invested monthly rather than at the end of the year. We shall use the approximation for the pilot study.

At the end of the second year, the employee will have approximately

$$P(2) = 2\beta S(2) + (1 + \gamma)P(1) = 2(0.0705)S(2) + 1.05(2820), \tag{1.1.2}$$

where

$$S(2) = (1 + \alpha)S(1) = 1.02S(1) = 1.02(20{,}000) = 20{,}400.$$

Thus, we have that $P(2) = 5837$. Similarly,

$$P(3) = 2\beta S(3) + (1 + \gamma)P(2) = 2(0.0705)S(3) + 1.05(5837), \tag{1.1.3}$$

where

$$S(3) = (1 + \alpha)S(2) = 1.02S(2) = 1.02(20{,}400) = 20{,}808.$$

So $P(3) = 9063$. By this point, we see how things are going well enough to leave the pilot study and set up the recurrence relations that solve the problem with general parameter values. Clearly, the key equations are

$$S(j + 1) = (1 + \alpha)S(j) \tag{1.1.4}$$

and

$$P(j + 1) = 2\beta S(j) + (1 + \gamma)P(j), \quad \text{for } j = 1, 2, \ldots, N_2 - N_1. \tag{1.1.5}$$

Moreover, at this point it is easy to take account of the fact that pension increments are paid monthly via

$$S(j + 1) = (1 + \alpha)S(j), \quad \text{for } j = 1, 2, \ldots, N_2 - N_1, \tag{1.1.6}$$

and

$$P_j(i + 1) = \frac{2\beta}{12} S(j) + \left(1 + \frac{\gamma}{12}\right) P_j(i), \quad \text{for } i = 0, 1, 2, \ldots, 11, \tag{1.1.7}$$

N_1 = starting age of employment
N_2 = retirement age
S = starting salary
P = starting principal
α = annual rate of increase of salary
β = fraction of salary contributed by employee
γ = annual rate of increase of principal in fund
Year = N_1
Month = 1
$$**P = \frac{\beta}{6}S + \left(1 + \frac{\gamma}{12}\right)P$$
Month = Month + 1
Is Month = 13?
If "no" go to **
If "yes" continue
Year = Year + 1
$S = S(1 + \alpha)$
Month = 1
Is Year = $N_2 + 1$?
Is "no" go to **
If "yes" continue
Return P

Figure 1.1.1. Subroutine annuity $(N_1, N_2, S, \alpha, \beta, \gamma)$.

where $P_{j+1}(0) = P_j(12)$. This system can readily be programmed on a handheld calculator or microprocessor using the simple flowchart in Figure 1.1.1. We find that the total stake of the employee at age 65 is a respectable 450,298 (recall that we have used an interest rate and a salary incrementation consistent with a low inflation economy).

We now know how much the employee will have in his pension account at the time of retirement. We wish to decide what the fair monthly payment will be if he invests the principal $P(N_2)$ in a life annuity. Let us suppose that actuarily he has a life expectancy of N_3 given that he retires at age N_2. To do this, we first compute the value to which his principal would grow by age N_3 if he simply invested it in an account paying at the prevailing interest of γ. But this is easily done by using the preceding routine with $S = 0$ and setting N_1 equal to the retirement age N_2 and N_2 equal to the expected time of death N_3. So we determine that using this strategy, the principal at the expected time of death is computed to be $P(N_3)$.

The monthly payments of the life annuity should be such that if they are immediately invested in an account paying at rate γ, then the total accrued principal at age N_3 will be $P(N_3)$. Let us suppose a guess as to this payment

X = guess as to fair monthly return
$P(N_2)$ = principal at retirement
γ = annual rate of return on principal
N_2 = retirement age
N_3 = expected age at death given survival until age N_2

$$P(N_3) = P(N_2)\left(1 + \frac{\gamma}{12}\right)^{N_3 - N_2}$$

**Call Annuity ($N_2, N_3, X, 0, 6, \gamma$)
Compare P with desired monthly payout
Make new guess for X and return to**

Figure 1.1.2. Program trial and error.

is X. Then we may determine the total principal at age N_3 by using the flowchart in Figure 1.1.1 using $S = X$, $\alpha = 0$, $\beta = 6$, $\gamma = \gamma$, N_1 = (retirement age)N_2, N_2 = (expected age at death)N_3. We can then find the fair value of monthly payment by trial and error using the program flowcharted in Figure 1.1.2.

We note that if the pensioner invests his principal at retirement into a fund paying 5% interest compounded monthly with all dividends reinvested, then at the expected death date he would have at that time a total principal of 741,650.

$$P(N_3) = 450{,}298\left(1 + \frac{0.05}{12}\right)^{(75-65)12} = 741{,}650. \tag{1.1.8}$$

Now on a monthly basis, the pensioner should receive an amount such that if he invested each month's payment at 5%, then at age 75 his stake would have increased from 0 to 741,650. As a first guess, let us try a monthly payout of 5000. Using the program in Figure 1.1.2, we find a stake of 776,414. Since this is a bit on the high side, we next try a payout rate of 4500—producing a stake at age 75 of 737,593. Flailing around in an eyeballing mode gets us to within one dollar of $P(N_3)$ in a number of iterations highly dependent on the intuition of the user. Our number of iterations was nine. The equitable monthly payout rate is 4777.

It should be noted in passing that our pensioner should be living rather well—far better than the case with the current system of Social Security which requires approximately the same rate of contribution as the private system considered above. Of course, the current Social Security System has other benefits which the employee could elect to have incorporated into his payout plan, for example, survivorship benefits to surviving spouse, Medicare, and disability. These additional add-ons would not cost sufficiently to lower the fair monthly payout below, say, 3000 per month. And we recall that these are dollars in a low inflation economy. Considering that there is some doubt that an individual entering the current job market will ever receive anything from Social Security at retirement, a certain amount of indignation on the part of the young is perhaps in order. Furthermore, the above private alternative to Social Security would allow the investments by the employee and his employer in the plan to be utilized as capital for investment in American industry, increasing employment as well as productivity.

The method of trial and error is perhaps the oldest of the general algorithmic approaches for problem solving. It is highly interactive; that is, the user makes a guess (in the above case X) as to the appropriate "input" (control variable, answer, etc.) which he feeds into a "black box." An output result is spewed out of the black box and compared to the desideratum—in the above problem, P and $P(N_3)$.

We note that the above example is one in which we know the workings of the black box rather well. That is, we have a model for reality that appears to be precise. And little wonder, for annuity is a manmade entity and one should be able to grasp the reality of its workings far better than, say, daily maximum temperatures in Houston forecast 3 years into the future.

In actuality, however, even the pension fund example is highly dependent on a series of questionable assumptions. For example, the interest figures used assume negligible inflation—a fair assumption for the mid-1980s but a terrible one for the late 1970s. Objections as to the assumption that the pension fund will be soundly managed are not too relevant, since most such funds are broadly invested, essentially a random selection from the market. Objections as to the uncertainty of employment of the employee in his current company are also irrelevant, since it is assumed that the vesting of such a fund is instantaneous, so that the employee loses no equity when he changes jobs. The plan suggested here is simply the kind of private IRA arrangement used so effectively by the Japanese both as a vehicle of retirement security and capital formation. Such a plan can reasonably be designed to track the performance of the overall economy. But the assumptions as to the specific yields of the plan will almost certainly be violated in practice.

At a later time, we shall cover the subject of *scenario analysis,* in which the investigator frankly admits he does not fully understand the black box's workings very well and examines a number of reasonable sets of scenarios (hypotheses) and sees what happens in each. At this point, we need only mention that in reality we are always in a scenario analysis situation. We always see through a glass darkly.

Having admitted that even in this idealized situation our model is only an approximation to reality, we observe that a wide variety of algorithms exist for solving the problem posed. Usually, if we can come up with a realistic mathematical axiomitization of the problem, we have done the most significant part of the work. The trial and error approach has many advantages and is not to be despised. However, its relative slowness may be somewhat inefficient. In the present problem, we are attempting to pick X so that P is close to $P(N_3)$. It is not hard to design an automated algorithm that behaves very much like the human mind for achieving this goal. For example, in Figure 1.1.3, we consider a plot of $G = P - P(N_3)$ versus X. Suppose that we have computed G for two values X_{n-1} and X_n. We may then use as our next guess X_{n+1}, the intercept on the X axis of the line joining $(X_{n-1}, G(X_{n-1}))$ and $(X_n, G(X_n))$.

Using the program in Figure 1.1.4, we use as our first guess a monthly output of 0, which naturally produces a stake at 75 years of age of 0. As our second guess, we use $X_2 = 450{,}298/(7 \times 12)$. With these two starting values of X_n and G_n, the program converges to a value that gives $P = 741{,}650$ to within one dollar in three iterations. The equitable payout rate obtained by the secant method is, of course, the same as that obtained by trial and error—namely, 4777.

1.2. INCOME TAX BRACKET CREEP

Beginning in 1981, there were a number of changes in the U.S. income tax laws. The major reason advanced for these modifications in the tax regulations was something called "bracket creep." This is a phenomenon of the progressive income tax which causes an individual whose income increases at the same rate as inflation to fall further and further behind as time progresses. There was much resistance on the part of many politicians to this indexing of taxes to the inflation rate, since the existing tax laws guaranteed a 1.6% increase in federal revenues for every 1% increase in the inflation rate. Another problem that was addressed by the tax changes in the early 1980s was the fact that a professional couple living together in the unmarried state typically paid a few thousand dollars less in taxes than if they were married. Those who felt the need for indexing and some relief from the "marriage tax" had carried out several "if this goes on" type scenario analyses. We consider one such below. All the figures below use typical salary rates for 1980 and an inflation rate a bit below that experienced at that time. The tax brackets are those of the 1980 IRS tables.

Let us consider the case of John Ricenik who accepts a position with a company which translates into a taxable income of $20,000. Let us project John's earning profile in the case where both inflation and his salary increase at an annual rate of 7%. First of all, we see that John's income will grow annually according to the formula

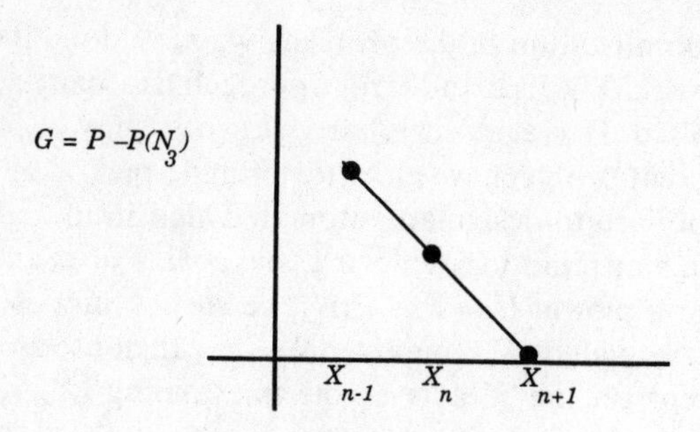

Figure 1.1.3. The secant method.

$P(N_2)$ = principal at retirement
γ = annual rate of return on principal
N_2 = retirement age
N_3 = expected age at death given survival until age N_2

$$P(N_3) = P(N_2)\left(1 + \frac{\gamma}{12}\right)^{12(N_3 - N_2)}$$

$$X_{n-1} = 0$$

$$G_{n-1} = -P(N_3)$$

$$X_n = \frac{P(N_3)}{12(N_3 - N_2)}$$

**Call Annuity $(N_2, N_3, X_n, 0, 6, \gamma)$

$$G_n = P - P(N_3)$$

$$\text{Slope} = \frac{G_n - G_{n-1}}{X_n - X_{n-1}}$$

$$X_{n+1} = X_n - \frac{G_n}{\text{Slope}}$$

Call Annuity $(N_2, N_3, X_{n+1}, 0, 6, \gamma)$
$G_{n+1} = P - P(N_3)$
Is $|G_{n+1}| < 1$?
If "yes" print X_{n+1} and stop
If "no" continue
$X_{n-1} = X_n$
$G_{n+1} = G_n$
$X_n = X_{n+1}$
$G_n = G_{n+1}$
Go to **

Figure 1.1.4. Program secant method.

Table 1.2.1. Tax Rates for Single Taxpayers

If taxable income is			
Not over $2,300...		0	
Over	But not over		of the amount over
$2,300	$3,400	14%	$2,300
$3,400	$4,400	$154 + 16%	$3,400
$4,400	$6,500	$314 + 18%	$4,400
$6,500	$8,500	$692 + 19%	$6,500
$8,500	$10,800	$1,072 + 21%	$8,500
$10,800	$12,900	$1,555 + 25%	$10,800
$12,900	$15,000	$2,059 + 26%	$12,900
$15,000	$18,200	$2,605 + 30%	$15,000
$18,200	$23,500	$3,565 + 34%	$18,200
$23,500	$28,800	$5,367 + 39%	$23,500
$28,800	$34,100	$7,434 + 44%	$28,800
$34,100	$41,500	$9,766 + 49%	$34,100
$41,500	$55,300	$13,392 + 55%	$41,500
$55,300	$81,800	$20,982 + 63%	$55,300
$81,800	$108,300	$37,677 + 68%	$81,800
$108,300	...	$55,697 + 70%	$108,300

$$\text{income} = 20{,}000\,(1.07)^{\text{year}-1980}. \tag{1.2.1}$$

The tax required to be paid in any given year is easily determined from Table 1.2.1. Inflation, on an annual basis, can be taken care of by expressing all after tax amounts in 1980 dollars according to the formula

$$\text{value (in 1980 dollars)} = \frac{\text{nominal amount}}{(1.07)^{\text{year}-1980}}. \tag{1.2.2}$$

To determine John's after tax profile, we need to examine the 1980 tax tables for single taxpayers.

We can readily compute the 6 year horizon table for John Ricenik's after tax income in 1980 dollars (Table 1.2.2).

We immediately note that John is not holding his own against inflation in spite of the fact that his salary is increasing at the same rate as inflation. This is due to the fact that his marginal increases in salary are being taxed at rates higher than the average rate for the total tax on his earnings. The purpose of indexing is to see that the boundaries for the rate changes increase at the annual rate of inflation.

Let us now investigate the "marriage tax." Suppose that John Ricenik marries his classmate, Mary Weenie, who has the same earnings projections

Table 1.2.2. After Tax Income

Year	Nominal	Tax	Nominal After Tax Income	After Tax Income (1980 Dollars)
1980	$20,000	$4,177	$15,823	$15,823
1981	$21,400	$4,653	$16,747	$15,651
1982	$22,898	$5,162	$17,736	$15,491
1983	$24,501	$5,757	$18,744	$15,300
1984	$26,216	$6,426	$19,790	$15,098
1985	$28,051	$7,142	$20,909	$14,908

as does John—that is, 7% growth in both salary increments and inflation. You might suppose that computing the after tax income of the Ricenik family is trivial. All one has to do is to double the figures in Table 1.2.2. This is, in fact, the case if John and Mary live together without being legally married.

There is another table that applies to John and Mary if they are legally husband and wife. We show this in Table 1.2.3.

In Table 1.2.4, we compare the tax John and Mary must pay if they are living in common law marriage as compared to that if they are legally married. All figures are given in nominal dollar amounts (i.e., in dollars uncorrected for inflation).

Table 1.2.3. 1980 Tax Schedule for Married Couples

Over	But not over		of the amount over
$3,400	$5,500	14%	$3,500
$5,500	$7,600	$294 + 16%	$5,500
$7,600	$11,900	$630 + 18%	$7,600
$11,900	$16,000	$1,404 + 21%	$11,900
$16,000	$20,200	$2,265 + 24%	$16,000
$20,200	$24,600	$3,273 + 28%	$20,200
$24,600	$29,900	$4,505 + 32%	$24,600
$29,900	$35,200	$6,201 + 37%	$29,900
$35,200	$45,800	$8,162 + 43%	$35,200
$45,800	$60,000	$12,720 + 49%	$45,800
$60,000	$85,600	$19,678 + 54%	$60,000
$85,600	$109,400	$33,502 + 59%	$85,600
$109,400	$162,400	$47,544 + 64%	$109,400
$162,400	$215,400	$81,464 + 68%	$162,400
$215,400	...	$117,504 + 70%	$215,400

Table 1.2.4. The Marriage Tax

Year	Combined Income	Tax if Married	Tax if Unmarried	Marriage Tax
1980	$40,000	$10,226	$8,354	$1,872
1981	$42,800	$11,430	$9,306	$2,124
1982	$45,796	$12,717	$10,324	$2,394
1983	$49,002	$14,289	$11,514	$2,775
1984	$52,432	$15,970	$12,852	$3,118
1985	$56,102	$17,768	$14,284	$3,484

We note that, even corrected for inflation, the marriage tax is increasing year by year. For example, when measured in constant dollars, the 1980 marriage tax of $1872 grows to $2484 in 1985.

In spite of the tax disadvantages, John and Mary decide to get married. Their living expenses, until they can buy a house, are $27,000/year (in 1980 dollars). To buy a house, the Riceniks need a down payment of $10,000 (in 1980 dollars). Assume that at the end of each year until they can make a down payment they invest their savings at an annual rate of 8% in short-term tax-free bonds. How many years must the Riceniks save in order to acquire their home if their wages and inflation increase at an annual rate of 7% and if the tax tables for 1980 had been kept in place? We answer this question by creating Table 1.2.5. As we note from the table, the Riceniks actually would never be able to afford their house under the conditions given. From 1985 on they would actually see the diminution of their savings.

Next, let us give a savings profile of the Riceniks with conditions as given above, except with the change that income tax levels are indexed by inflation, and the marriage penalty has been eliminated. Here, the relevant taxes in 1980

Table 1.2.5. The Savings Profile of John and Mary Ricenik

Year	Income	Tax	After Tax Income	Yearly Savings	Accumulated Savings	Accumulated Savings in 1980 Dollars
1980	$40,000	$10,226	$29,774	$2,274	$2,274	$2,274
1981	$42,800	$11,430	$31,370	$1,945	$4,401	$4,113
1982	$45,796	$12,718	$33,078	$1,593	$6,346	$5,543
1983	$49,001	$14,289	$34,712	$1,023	$7,877	$6,430
1984	$52,432	$15,970	$36,462	$415	$8,922	$6,807
1985	$56,102	$17,768	$38,334	–$236	$9,400	$6,702

Table 1.2.6. Savings Profile with Indexing and No Marriage Tax

Year	Taxable Income	Tax	Salary Income	Yearly Savings	Total Savings
1980	$40,000	$8,354	$31,646	$4,146	$4,146
1981	$40,000	$8,354	$31,646	$4,185	$8,331
1982	$40,000	$8,354	$31,646	$4,224	$12,555
1983	$40,000	$8,354	$31,646	$4,263	$16,818
1984	$40,000	$8,354	$31,646	$4,304	$21,122
1985	$40,000	$8,354	$31,646	$4,343	$25,465

dollars can be obtained by doubling $4177, the tax for a single person earning $20,000 per annum. This then gives us Table 1.2.6 (in 1980 dollars).

Thus, if the income tax were indexed to inflation, the Riceniks could afford the down payment on their home in less than 3 years even if their salaries only kept pace with inflation. In the above case, the only participation the Riceniks see in the assumed increase of productivity made possible by technology is through the interest on their capital. Even without this interest, they would be able to move into their home in 1982. It is perhaps interesting to note that most people who go through the computations in Table 1.2.5 do not believe the results when they create the table for the first time.

Next, let us briefly consider one of the consequences of the pre-1987 U.S. income tax—the tax shelter. The top tax level on salaried income was reduced in the early 1980s to 50%—a reform instituted during the Carter administration. However, it is not surprising that many people would search for some legal means of avoiding paying this "modest" rate of taxation. This can be achieved by noting that income which is gained by making an investment and selling it at least 6 months later is discountable in reporting total income by 60%. Thus, a property that is purchased for $100 and sold for $200 results in a profit of $100, but a taxable profit of only $40. Thus, the after tax profit is not $50 but $80. Not surprisingly, this "loophole" made possible a large tax avoidance industry. We shall see below an example of what should by all reason be deemed a bad real estate investment which turns out to result in approximately the same after tax profit as that obtained by a "good" conventional investment.

Ms. Brown is an engineer whose income level puts her in the 50% bracket on the upper level of her income tax. She decides to purchase a real estate acreage for a nominal price of $100,000. The terms are $10,000 down, and $10,000 payable at the beginning of each year with 10% interest on the balance payable 1 year in advance. Let us suppose Ms. Brown sells the property at the end of the fifth year. If the selling price is $135,000, how well has Ms. Brown

Table 1.2.7. A Real Estate Investment Outflow Profile

Year	Principal investment	Interest investment
1	$10,000	$9,000
2	$10,000	$8,000
3	$10,000	$7,000
4	$10,000	$6,000
5	$10,000	$5,000

done? Let us examine Ms. Brown's cash outflow during the 5 years (see Table 1.2.7).

Ms. Brown has invested $85,000 and still owes $50,000 on the property at the time of sale. Thus, it would appear that the result of five years' investment is that Ms. Brown gets back exactly what she has put into the investment. Furthermore, Ms. Brown must pay capital gains tax of 20% on the profit of $35,000. So, Ms. Brown's after tax stake at the end of the five years is $78,000. Clearly, Ms. Brown has not done well in her investing.

Let us suppose that Ms. Brown had rather followed a more traditional investment strategy. Suppose that she had taken the available salary income and invested in a money market fund at an annual rate of 10%. Now, we must note that the $9,000 interest payment which she made in the first year of the real estate investment is deductible from her gross income. If she had not paid it in interest, she would have had to pay half of it to the federal government in taxes. So she would not have had $19,000 to invest in a market fund in year 1—only $14,500. Moreover, the interest which she would earn in year 1, $1,450, would have been taxed at the 50% rate. Consequently, the after tax capital of Ms. Brown using the money market strategy would be given as shown in Table 1.2.8.

Note that Ms. Brown has approximately the same after tax capital using either the "imprudent" tax shelter or the "prudent" money market. It is not that the real estate investment has become good; it is rather that the tax system

Table 1.2.8. After Tax Principle Using Money Market Strategy

Year	After tax capital at year's end
1	$15,225
2	$30,686
3	$46,396
4	$62,365
5	$78,609

has made the money market investment bad. This is an example of the means whereby individuals have been forced by the tax system into making unproductive investments rather than depositing the money with lending agencies who would, in turn, lend out the money for capital development.

In September of 1986, the Rostenkowski-Packwood Tax Reform eliminated the capital gains preference, thus removing the incentive for the kind of bad investment mentioned here. Since much of the capital gains investment had favored the rich, Packwood and Rostenkowski set the marginal rate for families in the $150,000/year and up range at 28%. Marginal tax rates for the upper middle class, on the other hand, were set at 33%. So tax treatment which favors the wealthy and shafts the middle class has now been frankly and straightforwardly institutionalized instead of concealed with a capital gains exclusion. Other Rostenkowski–Packwood reforms included a suspension of the marriage deduction and that for state sales taxes. Moreover, the inflation indexing provisions of the reforms of the early 1980s appeared to be put at the hazard, "since the marginal rates are now so low." The "transitional" year marginal rates for the upper middle class of 38% were proposed to be made permanent by the Speaker of the House, Jim Wright. Some have cynically argued that, having wiped out a host of deductions in exchange for lower rates, the Congress will now gradually raise the new rates. Much discussion has taken place as to whether the new "simplified" code will hurt business. Apparently, one business that will not be hurt is that of the income tax preparation firms, since the new tax code runs to around 2000 pages.

One thing that has not been changed by any of the tax reforms of the last decade is the Social Security tax on income below roughly $40,000 per worker (a ceiling raised steadily by legislation passed years ago). This tax is approximately 7% from the employee and 7% from the employer, for a pooled total of slightly more than 14%. The self-employed pay roughly 10%. Many citizens pay more in Social Security tax than in income tax. And, of course, the monies paid by the worker in Social Security tax are also subject themselves to federal income tax. In addition to the two federal income taxes mentioned above, most states also levy income taxes—amounting to sums as high as 25% of the federal income tax. Furthermore, almost all states and municipalities levy sales taxes, real estate taxes, and so on. The proportion of an American citizen's income which goes to pay taxes is now several times that two centuries ago when taxes were a major cause of the American Revolution against Great Britain. Still, things could have been worse; the British taxation rates today are much more severe even than those in the United States.

We have gone through this section to demonstrate how a relatively simple and well-defined system such as the U.S. income tax is by no means easy to grasp until one goes through some computations with actual figures. When we leave the comfortable realm of well-axiomitized man-made systems and go into the generally very imperfectly understood systems of the social and natural sciences, we may well expect our difficulties of comprehension to increase dramatically.

1.3. RETIREMENT OF A MORTGAGE

One of the (few) advantages that Americans over 40 have over their younger counterparts is that they may have purchased houses when mortgage interest rates were low in comparison to present values. Another advantage is the fact that they purchased their houses for one-half to one-tenth of present values. That is their good news.

The bad news is the fact that in the event of tardy mortgage payment, a bank or a savings and loan company or whoever holds the paper is much more inclined to foreclose on a house worth $250,000 on which is owed $15,000 at 5.5% interest than on the same house on which is owed $175,000 at 15%. In the event of a massive series of bank failures, it is quite possible that an individual householder with $30,000 in FDIC insured liquid assets may nonetheless find himself unable to come up with a monthly mortgage payment of $500 (FDIC payoffs can take months or years). Consequently, it is not surprising that there is some motivation for the prudently paranoid to pay off their mortgage early.

Let us now consider the case of Mr. and Mrs. Jones who have 10 years to run on their mortgage. They now owe a principal of $15,000 on which they pay 5.5% interest on the outstanding balance. Suppose recent mortgage rates are in the 15% range. How much should the Joneses have to pay to retire the mortgage? The naive answer is "obviously $15,000." But is the mortgage really worth $15,000 to the bank who holds it? Obviously not. If the bank had the $15,000 in hand, they could lend it out at 15%. We wish to find the fair market value of the Jones mortgage.

One answer that might be proposed is that the fair market value is that quantity—say x—which, when compounded at 15%, will be worth the same as $15,000 compounded at 5.5%. This amount may easily be computed via

$$x(1.15)^{10} = 15{,}000(1.055)^{10} \tag{1.3.1}$$

or

$$x = 15.000\left(\frac{1.055}{1.15}\right)^{10} = 6333.$$

But this answer is also wrong, for we recall that under the terms of a mortgage the principal is retired over the entire term of the mortgage—not at the end. As the mortgage company receives the principal, it can loan it out at the prevailing mortgage rate—say 15%. The usual rule of principal retirement for older mortgages is that monthly payout is determined so that each installment of principal retired plus interest on unpaid balance is equal. Thus, the Joneses are to pay out their mortgage over 120 months according to the rule

$$\begin{aligned}(1.3.2)\qquad y &= p_1 + 15{,}000\frac{0.055}{12}\\ &= p_2 + (15{,}000 - p_1)\frac{0.055}{12}\\ &= p_3 + (15{,}000 - p_1 - p_2)\frac{0.055}{12}\\ &= \cdots = p_{120} + (15{,}000 - p_1 - p_2 - \cdots - p_{119})\frac{0.055}{12},\end{aligned}$$

where p_i is the principal retired on month i. By the 120th payment, all the \$15,000 will have been retired. Hence,

$$(1.3.3)\qquad \sum_{i=1}^{i=120} p_i = 15{,}000.$$

Using the fact that monthly payments are equal, we have

$$(1.3.4)\qquad p_1 + 15{,}000\left(\frac{0.055}{12}\right) = p_2 + (15{,}000 - p_1)\left(\frac{0.055}{12}\right).$$

This gives

$$(1.3.5)\qquad p_2 = \left(1 + \frac{0.055}{12}\right)p_1.$$

Similarly, from the second and third equalities in (1.3.2), we have

$$(1.3.6)\qquad p_3 = \left(1 + \frac{0.055}{12}\right)p_2.$$

or

$$(1.3.7)\qquad p_3 = rp_2,$$

where

$$r = 1 + \frac{0.055}{12}.$$

Thus, we have

$$p_i = r^{i-1} p_1. \tag{1.3.8}$$

Using the fact that $\sum p_i = 15{,}000$, we have

$$p_1(1 + r + r^2 + \cdots + r^{119}) = 15{,}000. \tag{1.3.9}$$

Recalling the formula for the sum of a geometric progression, we have

$$p_1\left(\frac{1 - r^{120}}{1 - r}\right) = 15{,}000. \tag{1.3.10}$$

This yields a p_1 value of \$94.04. Any other p_i value is immediately available via $p_i = r^{i-1}94.04$. The monthly payment is given simply by

$$94.04 + 15{,}000\left(\frac{0.055}{12}\right) = 162.79.$$

Moreover, the interest payment for the ith month is simply obtained via

$$\left(15{,}000 - \sum_{j=1}^{j=i-1} p_j\right)\frac{0.055}{12}. \tag{1.3.11}$$

But this is just

$$\left(15{,}000 - p_1 \sum_{j=0}^{j=i-1} r^{j-1}\right)\frac{0.055}{12} \tag{1.3.12}$$

or

$$\left(15{,}000 - p_1 \frac{1 - r^i}{1 - r}\right)\frac{0.055}{12}. \tag{1.3.13}$$

We are now in a position to answer the question as to the fair value for an instantaneous payment of the Jones mortgage. If the mortgage is paid out according to the terms of the original contract, the income realized by the mortgage holder on the first month is simply

$$p_1 + 15{,}000\frac{0.055}{12}. \tag{1.3.14}$$

Assuming the mortgage company lends the first month's payment at 15%, by the second month it will have realized a cumulative income of

$$p_2 + (15{,}000 - p_1)\frac{0.055}{12} + \left(p_1 + 15{,}000\frac{0.055}{12}\right)\left(1 + \frac{0.15}{12}\right). \tag{1.3.15}$$

By the third month, this will have grown to

$$p_3 + (15{,}000 - p_1 - p_2)\frac{0.055}{12} + \left[p_2 + (15{,}000 - p_1)\frac{0.055}{12} + \left(p_1 + 15{,}000\frac{0.055}{12}\right)\left(1 + \frac{0.15}{12}\right)\right]\left(1 + \frac{0.15}{12}\right). \tag{1.3.16}$$

Clearly, the above system is getting cumbersome. We need to replace the particular interest and capital values, and so on with algebraic symbols if we are to make sense of what is going on. The general recursion relation is easily seen to be

$$P_N = P_{N-1}s + (C - p_1 - p_2 - \cdots - p_{N-1})I \tag{1.3.17}$$

where

P_N = the company's capital at the end of the Nth month

$s = 1 + 0.15/12$

$I = 0.055/12$

$C = 15{,}000$

$p_i = p_1 r^{i-1}$

$r = 1 + 0.055/12.$

By simply using (1.3.17) recursively 120 times, we arrive at $P_{120} =$ \$44,802.42. This also gives us the ultimate solution for the equitable payoff by solving the equation

$$X_s^{120} = 44{,}802.42. \tag{1.3.18}$$

The solution is \$10,090.15. This is the amount the Joneses should expect to pay in order to retire mortgage immediately. As a matter of fact, the mortgage company is unlikely to settle for a penny less than the face value of \$15,000. Why? Because the mortgage is entered on the list of the company's assets as \$15,000. If lending institutions sold their old mortgages for their actual values, many would find themselves in the unenviable state of bankruptcy.

Had we been analyzing the above model before the advent of the electronic calculator, we would need to attempt to transform the recursion relation together with boundary conditions into a "closed form solution." That is, it would simply have been so troublesome to iterate the recursion formula 120 times that it would have been worth our while to go through the algebra required to bring things to a more manageable procedure for finding P_{120}. Just to give the reader a bit of feel for the relative cognitive utility of the recursion relation and the hazards of seeking a closed form solution, we shall demonstrate two methods for obtaining a closed form solution.

First, we could get lucky and observe that the value of the mortgage under existing terms to the mortgage holder is given simply by

$$162.79(1 + s + s^2 + \cdots + s^{119}) = 162.79\frac{1 - s^{120}}{1 - s} = 44{,}802.42. \tag{1.3.19}$$

This is a fine way to proceed. But in a practical problem solving situation, it is not always clear that we shall find an easy path. Consider below a typical "false trail." We again write out a few of the early P_N formulas to give us a feel for the solution.

$$\begin{aligned} P_1 &= p_1 + CI; \\ P_2 &= p_2 + CI - Ip_1 + s(p_1 + CI) = CI(1 + s) + p_2 + p_1(s - I); \\ P_3 &= p_3 + CI - p_1 I - p_2 I - p_3 I + sIC(1 + s) + Isp_2 + p_1 s(s - I) \\ &= CI(1 + s + s^2) + p_3 + p_2(s - I) + p_1(s^2 - Is - I). \end{aligned} \tag{1.3.20}$$

The formula for P_3 quickly shows us how to write down further P_i formulas more or less concisely. Thus, we have

$$\begin{aligned} P_4 = {} & CI(1 + s + s^2 + s^3) + p_4 + p_3(s - I) \\ & + p_2(s^2 - I - Is) + p_1(s^3 - Is^2 - Is - I). \end{aligned} \tag{1.3.21}$$

More generally then, we have

(1.3.22)

$$P_N = CI\sum_{j=0}^{j=N} s^j + \sum_{j=0}^{j=N-1} p_{j+1}s^{n-1-j} - I\sum_{j=1}^{j=N-1} p_j \sum_{k=0}^{k=N-1-j} s^k$$

$$= CI\frac{1-s^{N+1}}{1-s} + p_1\sum_{j=0}^{j=N-1} r^j s^{N-1-j} - Ip_1\sum_{j=1}^{N-1} r^{j-1}\frac{1-s^{N-j}}{1-s}$$

$$= CI\frac{1-s^{N+1}}{1-s} + \frac{1-(r/s)^N}{1-r/s}p_1 s^{N-1}\frac{1-s+I}{1-s} - \frac{Ip_1(1-r^N)/(1-r)}{1-s}.$$

This closed form solution does not give us a great deal of insight into our problem. Of course, it does give the right answer. Indeed, plugging in the values for C, I, r, s, p_1, and N, we come up with the value of $P_N =$ \$44,802.42. It is to be noted that the effort to obtain a closed form solution is frequently ill directed in the light of the general availability of swift and inexpensive computing devices. Certainly, such would be the case if we go down the false trail above. It is not reasonable to suppose that we shall quickly get on the right track for a simple closed form or indeed if there is such a simple form. And, of course, the path to get to (1.3.22) would take a fair amount of time, which we might not have. Very frequently we shall be well advised in a problem like the above in which a recursion relation is easily and immediately written down to stay with it instead of trying to bring it to a closed form.

1.4. SOME MATHEMATICAL DESCRIPTIONS OF THE MODEL OF MALTHUS

In this chapter on models of growth we began with compound interest because, historically, this seems to be the oldest of the quantified growth models. People have been lending out money at interest rates since the dawn of civilization. Since such financial transactions are essentially human constructs, it is not surprising that the modeling of such a process has been well understood for so long. In a sense, it is as though the process proceeds directly from an idealized model. Of course, even here, in what should be the most straightforward kind of modeling situation, there are many problems. Inflation rates are not generally predictable; wars and revolutions disrupt the orderly process of commerce; bankruptcy laws are created to protect debtors from their creditors, and so on. The point is that even in the most simple kind of modeling—that is, the modeling of processes based on idealized models—we are unable to describe precisely what is going on.

When it comes to modeling processes that do not proceed directly from a man-made model, we shall expect to be peering through even muddier water.

Still, as we shall see, the attempt to try and understand at least in part what is happening is well worth the effort. In fact, these attempts to conceptualize portions of the world around us have existed as long as humankind. It is our ability of the last 300 years to mathematize our conceptions and the commercial impetus to do so which have been largely responsible for the rapid scientific and technological progress characterizing this period. And it is this progress that has been responsible for the dramatic improvement in the material standard of living in those countries where this progress has been permitted.

Let us now consider the revolutionary work of the Reverend Thomas Malthus, *Essay on the Principle of Population.* This book, strictly speaking, was not as explicitly mathematical as one might have wished. Malthus used words rather than equations. His basic thesis is succinctly, if ambiguously, given by, "Population, when unchecked, increases in a geometrical ratio. Subsistence increases only in an arithmetical ratio." If the qualifying phrase "when unchecked" had been omitted, we could have summarized these two sentences very simply, using the symbols P for population and F for food by the two differential equations

$$\text{(1.4.1)} \qquad \frac{dP}{dt} = \alpha P \quad \text{and} \quad \frac{dF}{dt} = \beta.$$

The solutions to these equations are given simply by

$$\text{(1.4.2)} \qquad P(t) = P(0)e^{\alpha t} \quad \text{and} \quad F(t) = F(0) + \beta t.$$

The consequence would then be that the population was increasing at an exponential rate without any constraint from the more slowly growing food supply. But Malthus coupled the two processes, population and food ("subsistence"), with the use of the phrase "when unchecked." This simple qualifying phrase implies that a shortage of food checks the population growth. How should this checking effect be incorporated into our mathematical model? We might first decide that our constant α is not a constant at all but rather a function of F and P. To do so, however, we have to express population and food in similar units. This might be accomplished by using as one unit of food that amount required to sustain one person for a given unit of time. We then arrive at the coupled differential equation model

$$\text{(1.4.3)} \qquad \frac{dP}{dt} = \alpha(F,P)P \quad \text{and} \quad \frac{dF}{dt} = \beta,$$

where $\alpha(F,P)$ is read "α of F and P." Note that whereas F affects P, there is no effect of P on F. Accordingly, we may write

$$F(t) = F(0) + \beta t. \tag{1.4.4}$$

Suppose we argue that the growth rate per person is proportional to the difference between the food available and the food consumed. Then we have

$$\frac{1}{P}\frac{dP}{dt} = a(F(0) + \beta t - P). \tag{1.4.5}$$

In the case where the food supply is unchanging (i.e., $\beta = 0$), this is the *logistic* model of Verhulst (1844). A simple integration by parts for this special case gives

$$P = \frac{FP(0)e^{atF}}{P(0)e^{atF} + F - P(0)}. \tag{1.4.6}$$

Consequently, we shall refer to (1.4.5) as the *generalized logistic* model. The solution is given by

$$P(t) = \frac{P(0)\exp\left[a\left(F(0)t + \frac{1}{2}\beta t^2\right)\right]}{1 + aP(0)\int_0^t \exp\left[aF(0)\tau + \frac{1}{2}a\beta\tau^2\right]d\tau}. \tag{1.4.7}$$

In the present application, (1.4.7) gives some promise from a curve fitting standpoint. For example, for t large (1.4.7) is approximately given by

$$P(t) \approx F(0) + \beta t. \tag{1.4.8}$$

For a long established society, we might expect the Malthusian argument would yield a population that tracks the supposedly linear growth of the food supply. This is consistent with the population growth in England and Wales from 1800 to 1950 as shown in Figure 1.4.1. Using 1801 as the time origin, and using population in units of 1 million, we can fit the line by eye to obtain

$$P(t) \approx 6.4 + 0.2(t - 1801). \tag{1.4.9}$$

Naturally, there is little chance of determining a, since population records in Britain in the remote past, when population growth may have been less limited by the growth in the food supply, are not available. However, in the

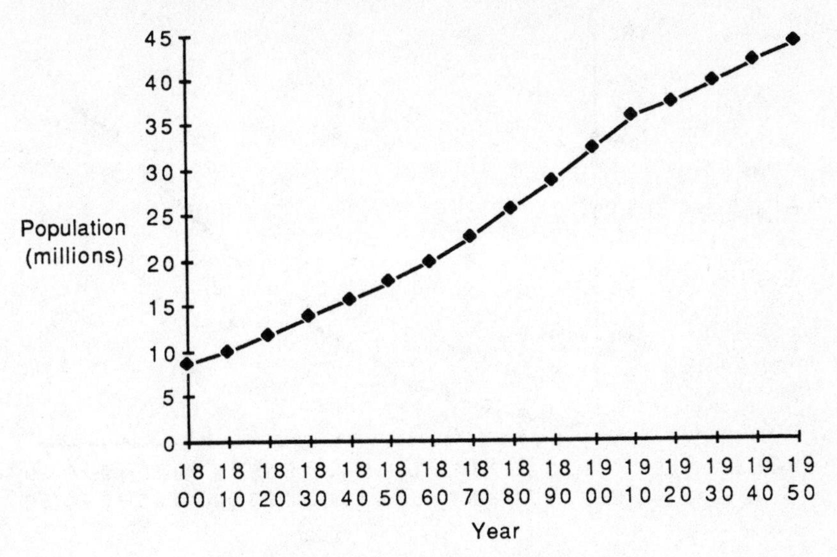

Figure 1.4.1. Population of England and Wales.

remote past, when t is "small," (1.4.7) becomes effectively

$$P(t) \approx P(0)e^{aF(0)t}. \tag{1.4.10}$$

Looking at the population growth of the United States in Figure 1.4.2, we note that whereas 20th century population growth is approximately linear, growth prior to, say, 1880 is much faster than linear. The only functional curve that can readily be identified by humans is the straight line. Accordingly, we take natural logarithms of both sides of (1.4.10). This gives us the equation of a straight line in t

$$log(P(t)) \approx \log(P(0)) + aF(0)t. \tag{1.4.11}$$

If (1.4.10) holds approximately over some time interval, then we should expect that a plot of $\log(P)$ versus time would yield very nearly a straight line. As we note from Figure 1.4.3, it seems that for the first 70 years of the 19th century, population growth is consistent with (1.4.10). We shall not at this point go throught the argument for estimating all the parameters in (1.4.5). But let us suppose that (1.4.5) could be made to fit closely both the American and British data. Would we have "validated" (1.4.5); that is, would we have satisfied ourselves that (1.4.5) is consistent with the data and our reason? Now a monotone data set can be fit with many curve families particularly if they have

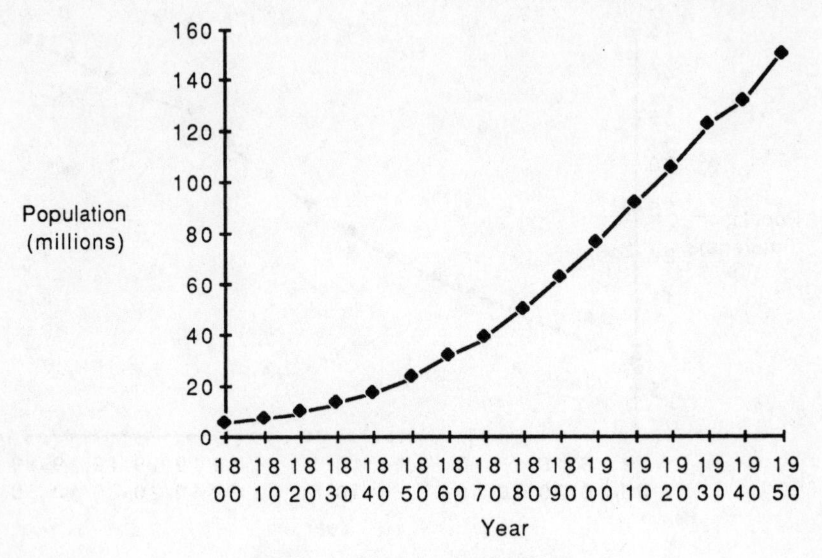

Figure 1.4.2. U.S. population growth.

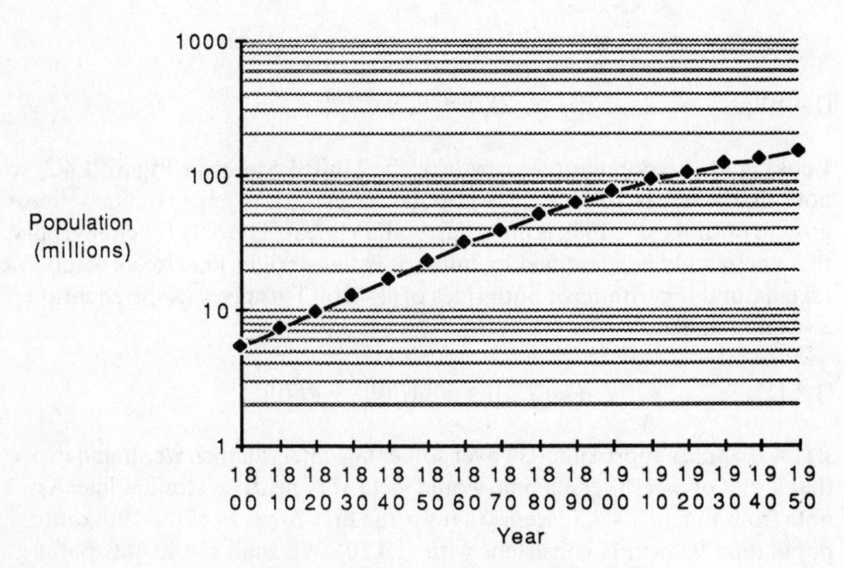

Figure 1.4.3. U.S. population growth.

three free parameters—as does (1.4.5). A good curve fit is a necessary condition but not a sufficient condition for establishing the "validity" ("plausibility" might be a better word) of a model.

Let us reexamine (1.4.5) in the light of the hypotheses of Malthus and our own reasoning. Is it reasonable to suppose that the per person population rate of increase will depend on the total availability of food? It would appear that if the food supply will support a population of 1,100,000 when the actual population is 1,000,000, then the per person rate of increase should be the same for a population of 100,000 which has an available food supply for 110,000. Accordingly, (1.4.5) is not appropriate. Rather, we should prefer

$$\frac{1}{P}\frac{dP}{dt} = \frac{a(F(0) + \beta t - P)}{P}. \tag{1.4.12}$$

The solution to this equation is

$$P(t) = \left(P(0) + \frac{\beta}{a} - F(0)\right)e^{-at} + F(0) - \frac{\beta}{a} + \beta t. \tag{1.4.13}$$

It is interesting to note that in the above, for large values of time, the population curve is given essentially by

$$P(t) = F(0) - \frac{\beta}{a} + \beta t. \tag{1.4.14}$$

Thus, the model (1.4.13) is consistent with the British population growth figures and with those from the United States after, say, 1910.

To examine the behavior of (1.4.13) for small values of t, we expand the exponential term neglecting all terms of order t^2 or higher (i.e., $O(t^2)$) to give

$$P(t) \approx \left[P(0) + \frac{\beta}{a} - F(0)\right](1 - at) + F(0) - \frac{\beta}{a} + \beta t; \tag{1.4.15}$$

that is,

$$P(t) \approx P(0) + [F(0) - P(0)]at. \tag{1.4.16}$$

Now we note that (1.4.13) departs from the spirit of Malthus, since growth consists essentially of two pieces linear in time with a transition in-between. The differential equation (1.4.12), moreover, has the problem that there is no "threshold" phenomenon insofar as per capita food supply is concerned. If

Table 1.4.1. U.S. Population 1800–1950

Date	t	$P(t)$ (in millions)
1800	0	5.308483
1810	10	7.239881
1820	20	9.638453
1830	30	12.866020
1840	40	17.069453
1850	50	23.191876
1860	60	31.443210
1870	70	38.558371
1880	80	50.155783
1890	90	62.947714
1900	100	75.994575
1910	110	91.972266
1920	120	105.710620
1930	130	122.785046
1940	140	132.169275
1950	150	150.697361

there is an excess capacity of food per person of 20, then people will multiply 20 times as fast as if the per capita excess capacity were 1. This is clearly unreasonable. However, if we fit the model and find out that the per capita excess food capacity is small throughout the extent of the data, then we might find the model satisfactory. Since it is not too much trouble to try a rough and ready fit, we shall attempt one, using U.S. data in Table 1.4.1.

First of all, using 1800 as the time origin, we have $P(0) = 5.308483$. Then, using the population figures for 1800 and 1810 as base points, we have from (1.4.16)

$$F(0) - \frac{0.19314}{a} = 5.308483. \tag{1.4.17}$$

Then, using (1.4.12) for $P(120)$ and $P(150)$, we have

$$\beta = 1.49956. \tag{1.4.18}$$

Finally, extending the line from $P(150)$ with slope 1.49956 back to the intercept on the population axis, we have from (1.4.12)

$$F(0) - \frac{1.49956}{a} = -74.9340. \tag{1.4.19}$$

Solving (1.4.17) and (1.4.19) together, we find

(1.4.20) $$F(0) = 17.1715$$

and

(1.4.21) $$a = 0.016281.$$

Already, we should have some concern about the size of $F(0)$. This leads to a per capita excess food supply in 1800 of around 2, probably an unreasonably high figure, and one that is high enough that the absence of a threshold effect in excess food supply might very well render our model less useful than one might hope.

Using these "rough and ready" parameter values in (1.4.13), we see in Figure 1.4.4 a comparison between the model values and the actual population figures. The quality of the fit is not spectacular. However, with three parameters to juggle—$F(0)$, a, *and* β—we could no doubt arrive at a very good fit with a little work. We might, for example, find the least squares solution by minimizing

(1.4.22) $$S(a, \beta, F(0)) = \sum_{j=0}^{j=15} \left[P(10j) - \left(P(0) + \frac{\beta}{a} - F(0) \right) e^{-10aj} - F(0) - \frac{\beta}{a} - 10j\beta \right]^2.$$

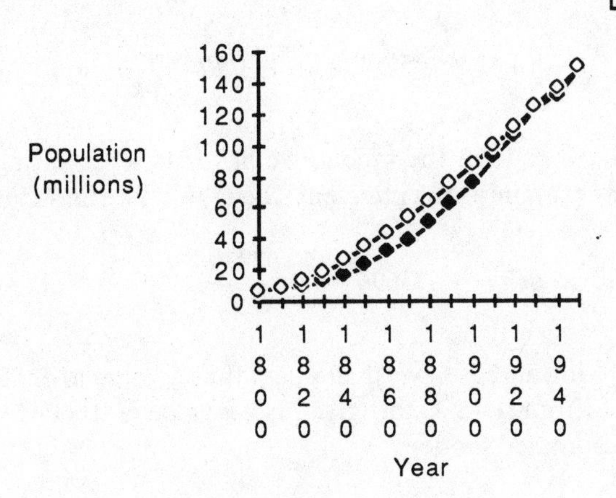

Figure 1.4.4. U.S. population growth and model values.

Now, let us go to a model more consistent with the apparent intent of Malthus. We need a model that will not penalize exponential growth when the excess per capita capacity of food is sufficiently large but will cause growth to be proportional to excess food capacity when the "affluence threshold" has been reached. There are an infinite number of models that will do this. It will be hard for us to choose among them. Consequently, we here borrow a maxim from the Radical Pragmatist William of Occam and try to pick a simple model that is consistent with Malthus' conjecture and with the facts before us. We suggest then

$$\frac{dP}{dt} = aP, \quad \text{when} \quad \frac{F(0) + \beta t - P}{P} > \frac{1}{k} \tag{1.4.23}$$

and

$$\frac{dP}{dt} = ak[F(0) + \beta t - P], \quad \text{otherwise.} \tag{1.4.24}$$

Examining the logarithmic curve in Figure 1.4.3, it appears that population growth is close to log–linear until around 1860. So we shall guess that the time where we switch from (1.4.23) to (1.4.24) is $t^* = 1860$. From the data at 1800 and 1860 we can estimate a via

$$a = \frac{1}{60} \ln\left(\frac{31.44321}{5.30843}\right) = 0.029648. \tag{1.4.25}$$

We know what the solution to (1.4.24) is from (1.4.13), namely,

$$P(t) = \left(P(t^*) + \frac{\beta}{ak} - F(t^*)\right) e^{-ak(t-t^*)} + F(0) + \beta t - \frac{\beta}{ak}. \tag{1.4.26}$$

Going to the essentially linear part of the model, we have, as when fitting (1.4.13), that $\beta = 1.49954$. By examining the intercept when $t = t^*$, we find that

$$F(t^*) = F(0) + \beta t^* = 15.73696 + \frac{50.57879}{k}. \tag{1.4.27}$$

We still must estimate $F(0)$. In this case, we shall guess that a value around 10 will be close to the truth. From (1.4.27), this gives us a k value of 0.600437. Finally, then, we have the estimated model

$$P(t) = P(0)e^{0.029648t}, \quad \text{before 1860,} \tag{1.4.28}$$

which will give us a new framework requiring that other data be collected, and so on.

If we take the position that the scientist should clear his or her mind of all prejudices about the generating mechanism of the data and "let the data speak for itself," we lose much. The human mind simply does not ever "start from zero" when analyzing phenomena. Rather, it draws on instinct plus a lifetime of experience and learning to which is added the current information. Empirical model building is simply a formulation of this natural learning process.

References

Malthus, Thomas R. (1798). *An Essay on the Principle of Population as It Affects the Future Improvement of Society, with Remarks on the Speculations of Mr. Godwain, M. Condorcet, and Other Writers,* London.

1.5. METASTASIS AND RESISTANCE

The killing effect of cancer is a result of the fact that a cancerous tumor grows, essentially, without limit and that it can spread to other parts of the body in a cascading sequence of *metastatic* events. If we can remove the tumor before it spreads via *metastatic progression,* that is, the breaking off of a cell from the primary which lodges in sites remote from it, then the chances of cure are excellent. On the other hand, once the primary has produced metastases to sites remote from the primary, surgical removal of the primary alone will generally not cure the patient's cancerous condition. If these metastases are widely spread, then, since each surgical intervention involves a weakening of the patient and the removal of possibly essential tissue, alternatives to surgery are required. Radiotherapy is also contraindicated in such cases, since radiation, like surgery, generally should be focused at a small region of tissue. Bartoszyński et al. (1982) have postulated that the tendency of a tumor to throw off a metastasis in a time interval $[t, t + \Delta t]$ is proportional to its size (in tumor cells) $n(t)$:

$$P(\text{metastasis in } [t, t + \Delta t]) = \mu n(t)\Delta t, \tag{1.5.1}$$

where μ represents the tendency to metastasize.

For all intents and purposes, only chemotherapy is left to assist a patient with diffuse spread of a solid tumor. Happily, we have a vast pharmacopoeia of chemotherapeutic agents that are very effective at killing cancerous tissue. Unfortunately, the general experience seems to be that within most detectable tumors, cells exist that resist a particular regime of chemotherapy. Originally, it was thought that such resistance developed as a feedback response of the malignant cells to develop strains that resisted the chemotherapy. Currently, it is believed that resistance develops randomly over time by mutation and that the eventual dominance of resistant cells in a tumor is simply the result of destruction of the nonresistant cells by the chemotherapeutic agent(s), a kind of survival of the fittest cells. It is postulated that, during any given time interval $[t, t + \Delta t]$,

$$P(\text{resistant cell produced during } [t, t + \Delta t]) = \beta n(t)\Delta t, \tag{1.5.2}$$

where β represents the tendency to develop irreversible drug resistance and $n(t)$ is the size of the tumor at time t.

Now if the resistant cells are confined to the primary tumor, a cure would result if the primary were excised and a chemotherapeutic agent were infused to kill the *nonresistant* cells which might have spread metastatically to other sites in the body. Unfortunately, it might very well happen that some resistant cells would have spread away from the primary or that originally nonresistant metastases would have developed by the time of the beginning of therapy.

Accordingly, it might be appropriate in some cases to follow surgery immediately by a regime of chemotherapy, even though no metastases have been discovered. The reason is that this kind of "preemptive strike" might kill unseen micrometastases which had not yet developed resistance to the chemotherapeutic agent(s). Such a strategy is termed *adjuvant chemotherapy*.

At the time of presentation, a patient with a solid tumor is in one of three fundamental states:

(1.5.3a) no metastases;

(1.5.3b) metastases, none of which contain resistant cells;

(1.5.3c) metastases, at least one of which contains resistant cells.

Both simple excision of the primary and the adjuvant regime of excision plus chemotherapy will cure a patient in state (1.5.3a). A patient in state (1.5.3c) will not be cured by either regime. Of special interest to us is the probability that, at the time of discovery (and removal) of the primary tumor, the patient is in state (1.5.3b). The probability that a patient has metastases, all nonresistant, at the time of presentation, gives us an indication as to the probability that a patient will be cured by adjuvant therapy but not by simple excision of the primary tumor.

For most types of solid tumors it is a close approximation to assume that the rate of growth is exponential, that is,

$$n(t) = e^{\alpha t}. \tag{1.5.4}$$

Since we may describe the time axis in arbitrary units, we lose no generality by making our primary tumor growth our "clock" and hence we may take α as equal to 1. Thus, for our purposes, we can use $n(t)$ as given simply by e^t, since for the parameter ranges considered, the amount of tumor mass removed from the primary to form the metastatic mass and/or the resistant mass is negligible (of relative mass, when compared to the primary of 1 part per 10,000). Furthermore, we assume that backward mutation from resistance to nonresistance is negligible.

Our task is to find a means of measuring the efficacy of adjuvant chemotherapy. We shall try to find estimates of the marginal improvement in the probability of cure of a solid tumor (with no apparent metastases) as a function of the size of the tumor and the parameters μ and β. We shall examine two approaches to this problem and suggest a third. We note that the problem suffers from the fact that the relatively simply Poisson axioms that describe the process go forward in time. On the other hand, to try to obtain the expression for marginal improvement in cure requires essentially a "backward" look. The first process will be an approximate one based on an argument used for a related problem of Goldie et al. (1982). First, let us look at the number of cells in resistant clones which develop in the primary tumor. This number, of course, is stochastic, but we shall approximate the number of resistant cells (R) by its expected value $E(R)$.

$$\frac{d(E(R))}{dt} \approx E(R) + \beta[e^t - E(R)] \tag{1.5.5}$$

has the solution

$$E(R) \approx e^t - e^{t(1-\beta)}. \tag{1.5.6}$$

Then

$$P(\text{no metastasis thrown off by resistant clone by } N) \approx \exp\left(-\mu(N-1) + \frac{\mu}{1-\beta}(N^{1-\beta} - 1)\right). \tag{1.5.7}$$

Similarly,

$$P(\text{no metastasis thrown off which develops resistant clone by } N) \approx \exp\left(-\beta(N-1) + \frac{\beta}{1-\mu}(N^{1-\mu} - 1)\right). \tag{1.5.8}$$

Summarizing,

(1.5.9) P(no resistant metastasis by total mass N)

$$\approx \exp\left(-(\beta + \mu)(N - 1) + \frac{\mu}{1 - \beta}(N^{1-\beta} - 1) + \frac{\beta}{1 - \mu}(N^{1-\mu} - 1)\right).$$

By differentiating (1.5.8) and (1.5.9) with respect to N, we note that for β very large relative to μ, the chances are that any formation of a resistant metastasis after tumor discovery would most likely be the result of spread from a resistant clone in the primary. Here, the standard protocol of removing the primary before beginning chemotherapy would be indicated. On the other hand, if the force of metastasis if much stronger than that of mutation to resistance, then chemotherapy might appropriately precede surgical intervention.

Now the approximation in (1.5.9) may be a serious underestimate for values of the stated probability below .25. The reason is that by using $E(R)$ instead of R, we are allowing clones of size less than one cell to form metastases, an obvious biological absurdity. Similarly, by replacing M by $E(M)$, we are allowing metastatic clones of less than one cell to develop and be at risk to resistance.

It is possible to obtain an exact expression for the marginal improvement possible from adjuvant chemotherapy, but only as the result of a fair amount of reflection. Let us consider the two events

(1.5.10) $A(N)$ = the event that by the time the total tumor mass equals N cells a nonresistant metastasis develops in which a resistant population subsequently develops (before a total tumor mass of N).

(1.5.11) $B(N)$ = the event that by the time the total tumor mass equals N cells a resistant clone develops from which a metastasis develops (before a total tumor mass of N).

We shall seek to compute $P(A^c(N))$ and $P(B^c(N))$. This can easily be computed by using (1.5.1) and (1.5.2).

(1.5.12)
P(metastasis occurs in $[t, t + \Delta t]$ followed by a resistant subclone before T)

$$= \mu e^t \left[1 - \exp\left(-\int_0^{T-t} \beta e^\tau \, d\tau\right)\right] \Delta t = \mu e^t [1 - e^\beta e^{-\beta N/n}] \Delta t$$

where $T = \ln(N)$. But then,

(1.5.13) $$P(A^c(N)) = \exp\left(-\int_0^T \mu e^t(1 - e^{\beta}e^{-\beta N/n})\,dt\right)$$
$$= \exp[\mu - \mu e^{\beta(1-N)} + \mu\beta e^{\beta}N\{Ei(-\beta) - Ei(-\beta N)\}].$$

Here, the *exponential integral* $Ei(\cdot)$ is defined for negative x by

(1.5.14) $$Ei(x) = -\int_{-x}^{\infty} \frac{e^{-t}}{t}\,dt.$$

Similarly, we obtain

(1.5.15) $$P(B^c(N)) = \exp[\beta - \beta e^{\mu}e^{-\mu N} + \mu\beta e^{\mu}N\{Ei(-\mu) - Ei(-\mu N)\}].$$

Thus, we can compute the probability that no resistant metastases have been formed by a total tumor size of N via

(1.5.16) $$P(\text{no resistant metastases}) = P(A^c(N))\,P(B^c(N)).$$

We have been able to obtain (1.5.13) and (1.5.15), exact solutions to the special two-stage branching processes defined by hypotheses (1.5.1)–(1.5.3) (generally deemed nontractable and requiring approximation), by exploiting the special structure of the case of interest to us here: namely, the nonappearance of any second-stage events. The exact expression in (1.5.16) is easily computed using, for example, the IMSL routine MMDEI for computing exponential integrals. However, for the magnitudes of N (very large) and μ and β (very small) in the present application, we obtain essentially the same result (using approximations for small arguments of $Ei(x)$) with

(1.5.17) $$P(\text{no resistant metastases by } N) = (\mu\beta\gamma^2)^{\mu\beta N}$$

where $\gamma = 1.781072$ is e raised to Euler's constant. The probability of no metastases at all by tumor size N is given simply by

(1.5.18) $$P(\text{no metastases by } N) = \exp\left(-\int_0^T \mu e^t\,dt\right) = e^{-\mu(N-1)}.$$

Thus, the probability a patient with a tumor of size N is curable by adjuvant therapy but not by simple excision of the primary tumor alone is given by

(1.5.19) $$P(\text{metastases, none of them resistant by } N)$$
$$= (\mu\beta\gamma^2)^{\mu\beta N} - \exp[-\mu(N-1)].$$

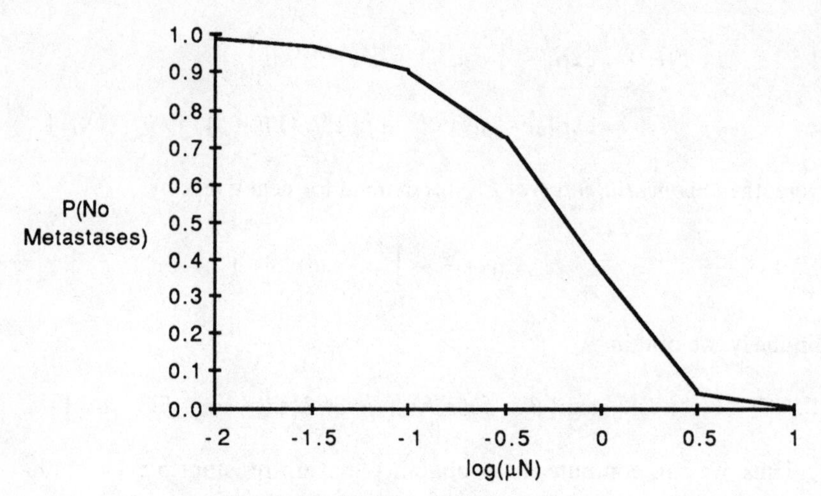

Figure 1.5.1. Nonoccurrence of metastases.

In Figure 1.5.1, we show the probability of nonoccurrence of metastases.

In Figure 1.5.2, we show the probability of nonoccurrence of resistant metastases versus $\log(N)$ for various values of $\log(\mu\beta)$.

Typical values for β for mammalian cells are 10^{-6} to 10^{-4}. A typical tumor size at time of detection is 10^{10} cells (roughly 10 cubic centimeters).

In Figure 1.5.3, we show the probability there exist metastases but no

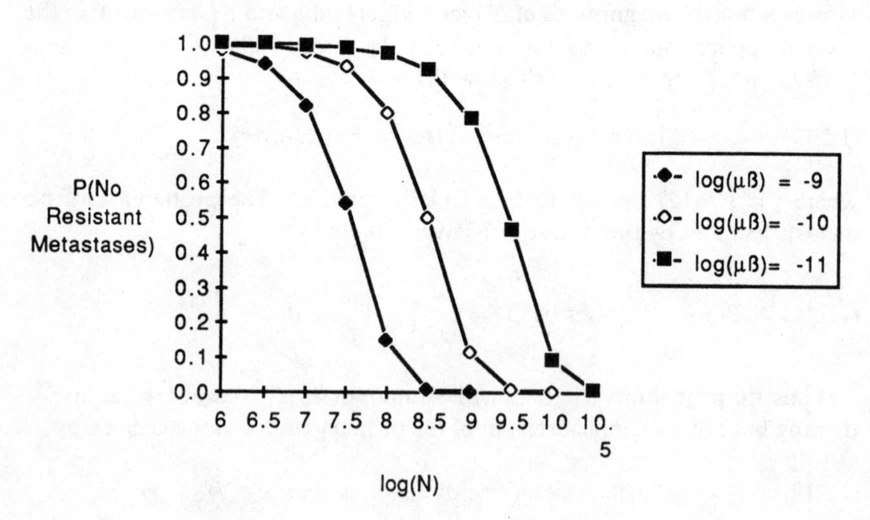

Figure 1.5.2. Nonoccurrence of resistant metastases.

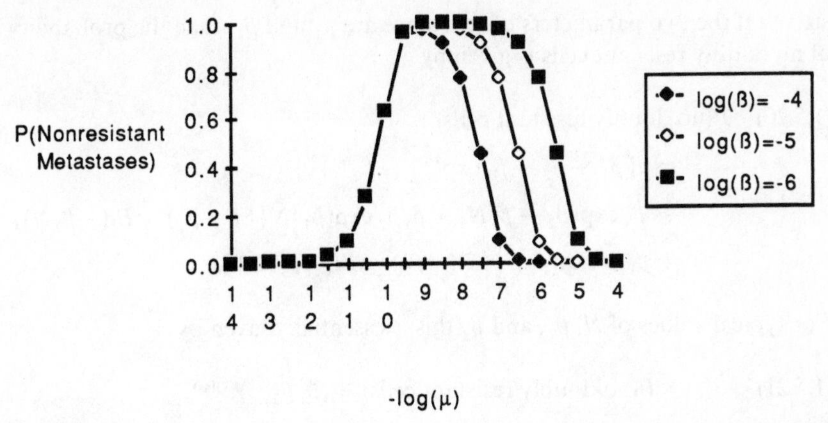

Figure 1.5.3. Probability of metastases but no resistant metastases when log(tumor cell count) = 10.

resistant metastases at a tumor mass of 10^{10} for various β and μ values. We note, for example, that for $\beta = 10^{-6.5}$ to 10^{-10}, the probability a patient presents with a condition curable by adjuvant therapy but not by simple excision of the primary is at least 40%. Similar results hold for early detection (10^9 cells) as shown in Figure 1.5.4. This wide range of μ values includes that reported for breast cancer by Bartoszyński et al. (1982).

As an aside, it should be noted that the argument used to obtain eqs. (1.5.13), (1.5.15), and (1.5.16) can be used for the situation where one is interested in obtaining the probability that resistance to both of two independent chemotherapeutic agents will not have developed by tumor mass of

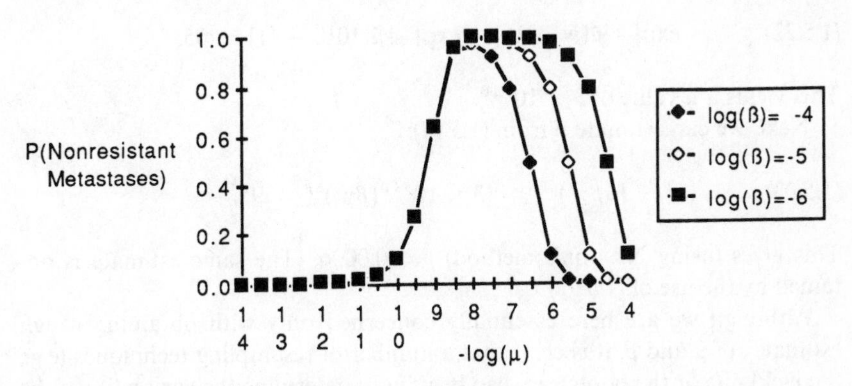

Figure 1.5.4. Probability there exist metastases but no resistant metastases when log(tumor cell count) = 9.

size N. If the two parameters of resistance are β_1 and β_2, then the probability of no doubly resistant cells is given by

(1.5.20) $$\begin{aligned} P(\text{no doubly resistant cells}) &= \exp[\beta_1 + \beta_2 - \beta_1 \exp(\beta_2 - \beta_2 N) \\ &\quad - \beta_2 \exp(\beta_1 - \beta_1 N) + \beta_1 \beta_2 \exp(\beta_1) N\{Ei(-\beta_1) - Ei(-\beta_1 N)\} \\ &\quad + \beta_1 \beta_2 \exp(\beta_2) N\{Ei(-\beta_2) - Ei(-\beta_2 N)\}]. \end{aligned}$$

For typical values of N, β_1, and β_2 this is essentially given by

(1.5.21) $$P(\text{no doubly resistant cells}) = (\beta_1 \beta_2 \gamma^2)^{\beta_1 \beta_2 N}.$$

In the case of breast cancer that has metastasized at least to local nodes, it has been reported by Buzdar et al. (1986) that the use of adjuvant chemotherapy decreases disease mortality by as much as 54% when compared to surgery alone. In order to estimate μ and β clinically, we need randomized trials on tumors (which have exhibited no metastases at presentation) using surgery followed by adjuvant chemotherapy compared with surgery alone. Such a clinical data base is not currently available. However, since we are, at this stage, really seeking rough estimates of μ and β, animal experiments may be appropriate.

We examine below how such experiments might be used to estimate μ and β. Let us suppose we have stratified our data by tumor size at presentation. Consider the 10^{10} primary cell stratum. Suppose that the control group (surgical excision only) exhibits a cure rate of 5% and that the adjuvant therapy group exhibits a 95% cure rate. Then we can estimate μ from (1.5.18):

(1.5.22) $$\exp[-\hat{\mu}(N - 1)] = \exp[-\hat{\mu}(10^{10} - 1)] = .05.$$

This yields a $\hat{\mu}$ value of 3×10^{-10}.

Next, we can estimate β from (1.5.17):

(1.5.23) $$(\hat{\mu}\hat{\beta}\gamma^2)^{\hat{\mu}\hat{\beta}N} = (3 \times 10^{-10}(\hat{\beta}\gamma^2)^{3\hat{\beta}} = .95.$$

This gives (using Newton's method) $\hat{\beta} = 0.0006$. (The same estimate is obtained by the use of (1.5.9).)

Although we are here essentially concerned only with obtaining rough estimates of μ and β, it is clear that a number of resampling techniques (e.g., the jackknife or the bootstrap) can be used to determine the variability of the estimates. Let us suppose, for example, that we have N_1 individuals in the control group of whom n_1 are cured and N_2 individuals in the adjuvant group of whom n_2 are cured. Using the coding of 1 for a cure and 0 for a noncure, we repeatedly sample (say, M times, where M is very large) N_1 individuals with probability of success n_1/N_1 and N_2 individuals with probability of success n_2/N_2. For each such jth sampling we obtain $\hat{\mu}_j$ and $\hat{\beta}_j$ as above. Then we have as ready bootstrap estimates for $\mathrm{Var}(\hat{\mu})$, $\mathrm{Var}(\hat{\beta})$, and $\mathrm{Cov}(\hat{\mu}, \hat{\beta})$, $\Sigma(\hat{\mu}_j - \bar{\mu})^2/M$, $\Sigma(\hat{\beta}_j - \bar{\beta})^2/M$, and $\Sigma(\hat{\mu}_j - \bar{\mu})(\hat{\beta}_j - \bar{\beta})/M$, respectively. An unacceptably large 95% confidence ellipsoid would cause us to question the validity of our model, and the comparative suitability of competing models might be judged by examining the "tightness" of the 95% confidence ellipsoids of each using the same data set.

In our discussion, note that both (1.5.9) and (1.5.17) are easy to compute. Equation (1.5.9) is much easier to think out than (1.5.17), but if we use (1.5.9), we know we are employing an approximation whose imprecision is hard to assess unless we have the more precise formula (1.5.17), which is time consuming to derive. It would clearly be good to have the advantage of the accuracy of (1.5.17) without the necessity for much effort spent in cogitation. We shall show how we can employ simulation to go in the forward direction, pointed to by the axioms in Section 3.3.

References

Bartoszyński, Robert, Brown, Barry W., and Thompson, James R. (1982). Metastatic and systemic factors in neoplastic progression, in *Probability Models and Cancer,* Lucien LeCam, and Jerzy Neyman, eds., North Holland, Amsterdam, pp. 253–264, 283–285.

Buzdar, A. U., Hortobagyi, G. N., Marcus, C. E., Smith, T. L., Martin, R., and Gehan, E. A. (1986). Results of adjuvant chemotherapy trials in breast cancer at M.D. Anderson Hospital and Tumor Institute, *NCI Monographs,* **1,** 81–85.

Efron, Bradley (1979). Bootstrap methods—another look at the jacknife, *Annals of Statistics,* **7,** 1–26.

Goldie, J. H. and Coldman, A. J. (1979). A mathematical model for relating the drug sensitivity of tumors to their spontaneous mutation rate, *Cancer Treatment Reports,* **63,** 1727–1733.

Goldie, J. H., Coldman, A. J., and Gudauskas, G. A. (1982). Rationale for the use of alternating non-cross-resistant chemotherapy, *Cancer Treatment Reports,* **66,** 439–449.

IMSL, Inc. (1985). Subroutine MMDEI.

Jahnke, Eugene and Emde, Fritz (1945). *Tables of Functions,* Dover, New York, pp. 1–3.

Thompson, James R. and Brown, Barry W. (1986). A stochastic model providing a rationale for adjuvant chemotherapy, Rice Technical Report 86-19, Department of Statistics.

CHAPTER TWO

Models of Competition, Combat, and Epidemic

2.1. AN ANALYSIS OF THE DEMOGRAPHICS OF ANCIENT ISRAEL BASED ON FIGURES IN THE BOOKS OF NUMBERS, JUDGES, AND II SAMUEL

Let us consider a simple model of population growth:

$$\frac{dY}{dt} = \alpha Y, \tag{2.1.1}$$

where Y is the size of the population at time t and α is the (constant) rate of growth. The solution is simply

$$Y(t) = Y_0 e^{\alpha t}, \tag{2.1.2}$$

where Y_0 is the size at $t = 0$. We note first that this model has several natural reasons for its appeal.

1. The model follows rather directly from the microaxiom:

   $$P(\text{individual gives birth in } [t, t + \Delta t]) = \alpha\,\Delta t. \tag{2.1.3}$$

   The expected increase in the interval is given simply by multiplying by Y. As Y becomes large, we can replace Y by its expectation to obtain

   $$\Delta Y = Y\alpha\,\Delta t. \tag{2.1.4}$$

2. The model has the happy property that if we choose to divide the Y population into subgroups, which we then watch grow, when we recombine them, we get the same result as when we use pooled Y all along.

Clearly, there are bad properties of the model, too. For example:

1. The model is more appropriate for single-cell animals than for people; there is no child-bearing age range considered in the model, no allowance for the fact that only women have children, and so on. However, for large samples, this objection is academic. We could use the number of men (as we shall of necessity do due to the presentation of our data) assuming rough proportionality between men and women. Age stratification will also not make a noticeable difference for large populations.
2. There is no factor for diminishing growth as the people fill up the living space. However, for the population to be considered here, it would appear that death by armed violence was a much more important factor than a limit to growth factor (although it can surely be argued that death in war was a function of "crowding").
3. There is no factor for death by war. We shall add such a factor to our model shortly.

Around 1700 B.C., there were 12 male Jews fathering children. These were, of course, the sons of Jacob, his two wives (Leah and Rachel), and two concubines (Bilhah and Zilpah). The entire family disappeared into Egypt shortly thereafter. Before the conquest of Canaan, in 1452 B.C., Moses conducted a census of the total military force of the Jews. The results are shown in Table 2.1.1. Since the Levites are always excluded from censuses, we must inflate the figure by 12/11 if we are to come up with a figure that can be used to estimate α. This gives us a comparable total of 656,433 and an estimate of

Table 2.1.1. Moses' Census of 1452 B.C.

Tribe	Number of Warriors
Judah	74,600
Issachar	54,400
Zebulun	57,400
Reuben	46,500
Simeon	59,300
Gad	45,650
Ephraim } Joseph	40,500
Manasseh } Joseph	32,200
Benjamin	35,400
Dan	62,700
Asher	41,500
Naphtali	53,400
Total	601,730

$\alpha = \ln(656{,}433/12)/248 = 0.044$. To put this in perspective, this growth rate is roughly that of Kenya today, and Kenya has one of the fastest growing populations in the world. Certainly, it was an impressive figure by Egyptian standards and is given as the reason (Exodus 1) for the subsequent attempts (using such devices as infanticide by midwives) of a few later Pharaohs to suppress the growth in Jewish population. We note that this growth rate would appear to indicate that the bondage of Israel was of relatively short duration and that, on the whole, the Jews had lived rather well during most of their time in Egypt. This is also consistent with later writings.

In Deuteronomy 23, there is an injunction to treat the Egyptians (along with the Edomites, descendants of Jacob's brother Esau) relatively well. They were allowed to become full partitipants in all forms of Jewish life after only three generations. (Contrast this with the Moabites who were not allowed membership in the congregation under any circumstances "even to their tenth generation." Obviously, the harsh Mosaic law was not always strictly observed, since King David and King Solomon had the Moabitess Ruth as their great grandmother and great great grandmother, respectively.) Also, the Jews, having gained control of Canaan, were very much inclined to enter into alliances with the Egyptians against their enemies to the north and east. The present day guarded friendship between Egypt and Israel has ancient precedent.

Now had Moses had a Macintosh, he could have represented this data graphically, using a pie chart as shown in Figure 2.1.1. We note that a pie chart is somewhat appropriate, since the circular representation does not give any notion of ordering of the attributes of tribeship (i.e., it does not demand

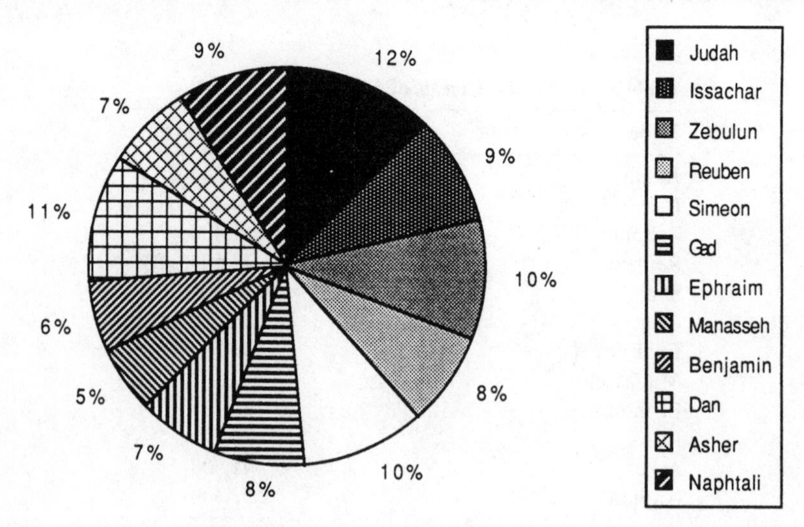

Figure 2.1.1. Israeli defense forces (1452 B.C.).

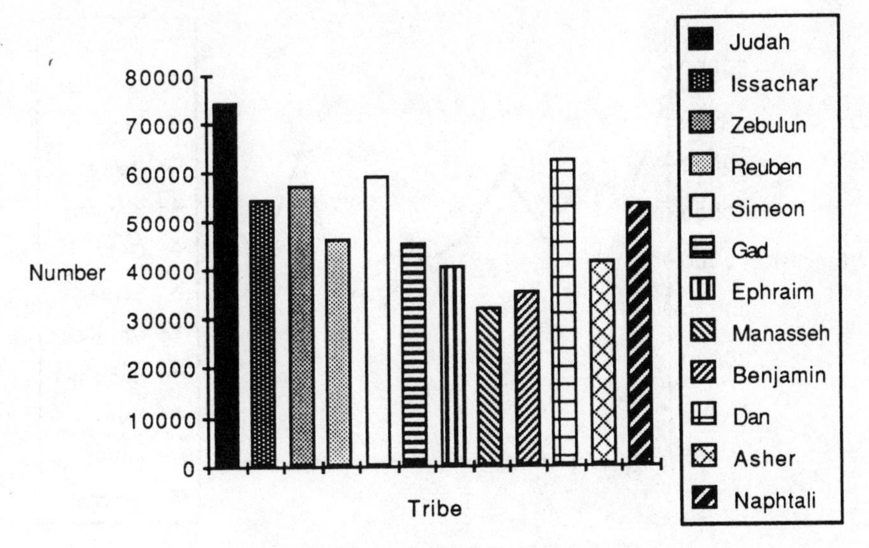

Figure 2.1.2. Israeli defense forces (1452 B.C.).

that we say that Issachar > Dan). It does give false notions of adjacency; for example, Asher is "between" Dan and Naphtali for no good reason. Less appropriate would have been a column chart (Figure 2.1.2), which does "order" one tribe versus another for no good reason.

Still less appropriate would have been a line chart (Figure 2.1.3), which presents information more misleadingly and less informatively than the other two. (Note, however, that any one of the charts could have been constructed from either of the others.)

In truth, for the present data set, we may be better off to stick with a table, since the graph may encourage us to make subconscious inferences that are accidents of presentation rather than the result of intrinsic attributes of the data.

Returning to our attempt to make some modeling sense of the data at hand, we note (Judges 20) that after the conquest of Canaan in 1406 B.C. we have more population information at our disposal. The total warrior population had declined to 427,000. Putting in the correction to account for the Levites, we come up with a total of 465,000. Recall that before the conquest, the comparable figure was 656,000. It is obvious now that we need to include a term to account for the losses in warfare during this period of 46 years:

$$\frac{dY}{dt} = \alpha Y - \lambda, \tag{2.1.5}$$

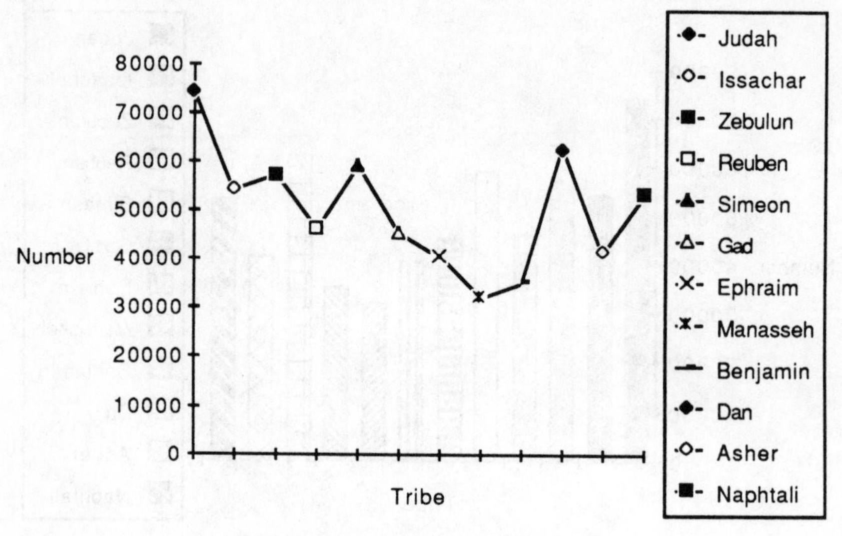

Figure 2.1.3. Israeli defense forces (1452 B.C.).

where λ is the annual loss to the male population due to warfare. The solution is simply

$$Y(t) = \frac{\lambda}{\alpha} + \left(Y_0 - \frac{\lambda}{\alpha}\right)e^{\alpha t}. \tag{2.1.6}$$

Now the figures for the two base years of 1452 B.C. and 1406 B.C. give us a means of estimating λ via

$$\lambda = \frac{\alpha}{e^{\alpha t} - 1}(Y_0 e^{\alpha t} - Y), \tag{2.1.7}$$

where the time origin is taken to be 1452 B.C.

If we use our previous estimate of $\alpha = 0.044$, we find $\lambda = 30{,}144$, giving total losses of males during the 46 year period of 1,386,606. Clearly, the conquest of Canaan had caused heavy casualties among the Jews. Moreover, total war was pursued, with the Jews killing everyone: men, women, and children. Assuming their opponents retaliated in kind, the above figure might be close to 3,000,000.

Now unsettled times can be expected to lower fecundity, so perhaps the use of $\alpha = 0.044$ is unrealistic. We do have a means of estimating the growth rate between 1406 and 1017, since we have another census at that later date (the warrior age male population exclusive of Levites was 1,300,000). Unfortunately, there was intermittent, frequently very intense, warfare with the Philistines, Midianites, Ammonites, and so on, during that period, so a new value of λ is really needed. But if we swallow the later warfare casualties in α, we come up with an estimate of 0.0029. If we use this value in our equation for estimating the total casualties during the conquest of Canaan, we come up with a λ of 5777 and a total warrior age male loss figure of 265,760. This figure is no doubt too small. But translated to U.S. population values, if we go between the two extreme figures, we should be talking about losses at least as great as 30,000,000.

One observation that ought to be made here is that in 1017 B.C. David had an army of 1,300,000 after 450 years of war, during which great sacrifices had been made and borne by the Jews. The period of peace that started about that year continued essentially until 975 B.C. After that date, the Kingdom was generally in subjugation and vassalage to some other state. What happened? Were the Jews overwhelmed by armies so massive that they could not match them?

To answer this question, we might consider some sizes of the largest field armies of the ancient world: see Table 2.1.2.

The notion of Israel as a numerically tiny nation, unable to cope with attacks by massive invading armies is seen to be false. At the time of the Kingdom of David (1017 till 975), Israel had a larger population than Rome had 700 years later. We know, not only from Biblical accounts but also from other records, that the Jews were one of the largest "national" groups in the ancient world. Their militia was vast (1,300,000—larger than the maximum size (750,000) ever attained by the total military forces of the Romans). What then caused the decline (which set in immediately after Solomon)? The short answer is *high taxes.* Whether these taxes were mainly the result of costs of the Temple of Solomon or the costs for the luxuriant life-style of Solomon, we cannot answer definitively. However, we know that in 975 B.C. a delegation pleaded with Solomon's son Rehoboam to lower the rate of taxation. While acknowledging that taxes had been high under his father ("My father has chastised you with whips...") Rehoboam promised little in the way of tax

Table 2.1.2. Field Armies of the Ancient World

Hannibal's army at Cannae	32,000
Largest Assyrian army	200,000
Xerxes' army in Greece	200,000
Alexander's army at Issus	30,000
Scipio's army at Zama	43,000
Rome's total forces under Augustus	300,000

relief ("But I will chastise you with scorpions."). Upon reception of this news, the larger part of the Jews seceded from the Kingdom of Judah, setting up the Kingdom of Israel. And within these kingdoms, there then began further quarrels and divisions. Frequently, the two kingdoms would go to war with each other, inviting help from outside powers. From the time of the death of Solomon, the Kingdom of David was on a slippery slope to destruction. The total time span of the Kingdom was a scant 42 years.

References

Gabriel, Richard A. (1984). *Operation Peace for Galilee,* Hill and Wang, New York.

2.2. THE PLAGUE AND JOHN GRAUNT'S LIFE TABLE

In A.D. 1662, some 3113 years after the census of Moses, an obscure haberdasher, late a captain in the loyalist army of Charles II, published an analysis on data originally collected by Thomas Cromwell, 127 years earlier, dealing with age at the time of death in London. The data had been collected at the request of the merchants of London who were carrying out the most basic kind of marketing research; that is whether potential customers (i.e., live people) were on the increase or decrease. Interestingly enough, the question originally arose because of the fact that the bubonic plague had been endemic in England for many years. At times, there would be an increase of the incidence of the disease, at other times a decrease. It was a matter of sufficient importance to be attended to by Chancellor Cromwell (also Master of the Rolls). Without any central data bank, a merchant might put a shop in an area where the decline in population had eliminated any potential opportunity, due to market saturation.

Cromwell's data base consisted in records of births and deaths from the Church of England to be carried out and centrally located by the clergy. Before Graunt, all analyses of the data had suffered the usual "can't see the forest for the trees" difficulty.

Graunt solved this problem and started modern statistics by creating Table 2.2.1.

Graunt also tabulated $1 - F$, where F is the cumulative distribution function. We can easily use this information to graph F (see Figure 2.2.1).

From Graunt's table, it is an easy matter to compute the life expectancy—18 years. It would appear the plague was in full force. Note that Graunt's brilliant insight to order his data made possible a piece of information simply not available to the ancient Israeli statisticians, the sample average. How

Table 2.2.1. Graunt's Life Table

Age Interval	P(death in interval)
0–6	.36
6–16	.24
16–26	.15
26–36	.09
36–46	.06
46–56	.04
56–66	.03
66–76	.02
76–86	.01

obvious Graunt's step seems to us today. Yet, it would appear he was the first to take it. It is also interesting to note how application frequently precedes theory. Graunt's sample cumulative distribution function predates any notion of a theoretical cdf.

Following our earlier graphical analyses, we note that the pie chart (Figure 2.2.2) is rather less informative than the bar chart (Figure 2.2.3, near histogram), which is slightly less useful than the near histopolygon (Figure 2.2.4, line chart). The pie chart draws the viewer's attention to a periodicity which simply does not exist. If we divide the probabilities by the width of the age interval, we could get a true histogram as shown in Figure 2.2.5 (with unequal intervals).

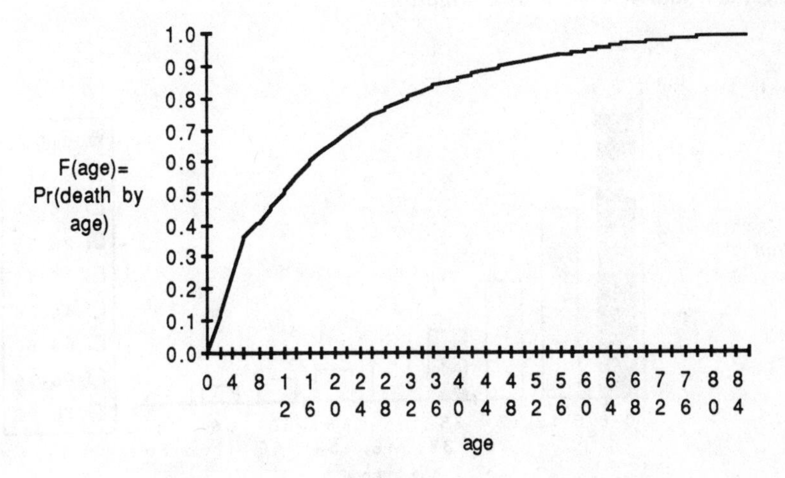

Figure 2.2.1. Cumulative sum of Graunt's histogram.

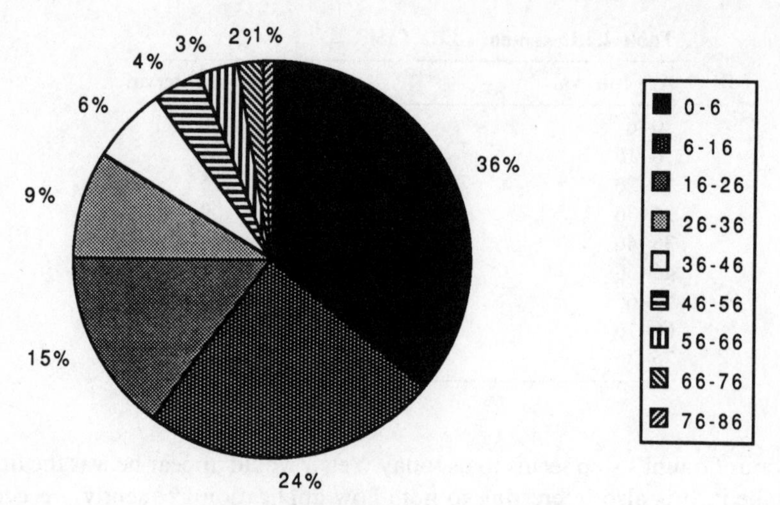

Figure 2.2.2. Graunt's life table (A.D. 1661).

We observe that Graunt's table cries out to be graphed, as the demographic data from ancient Israel did not. What is the difference? In the Israeli data, there is no natural measure of proximity of tribal attributes. The covariate information is completely qualitative. Dan cannot be said, in general, to be "closer" to Napthtali than to Benjamin. A 5 year old Englishman is very like a 7 year old and very different from a 70 year old. Graunt had empirically discovered a practical realization of the real number system before the real numbers were well understood. In so doing, he also presented the world with its first cumulative distribution function.

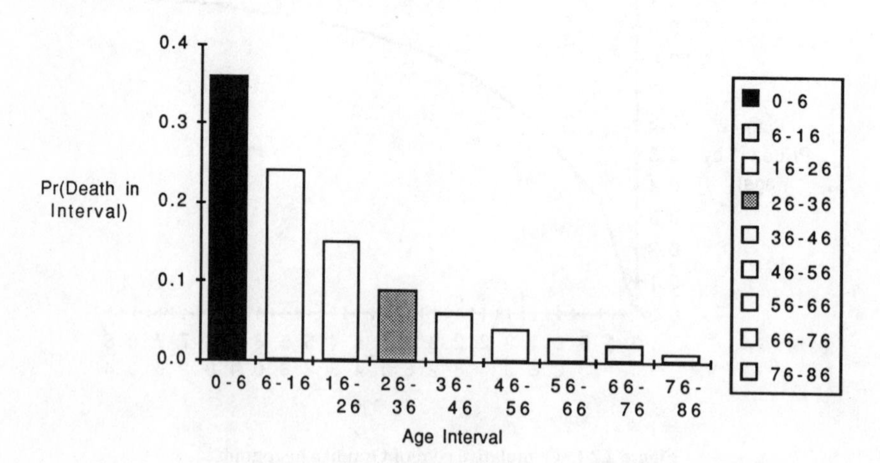

Figure 2.2.3. Graunt's life table (A.D. 1661).

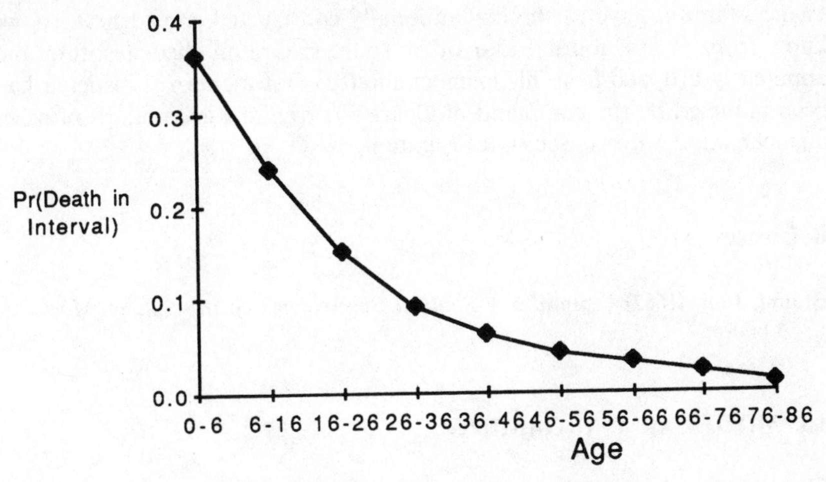

Figure 2.2.4. Graunt's life table (A.D. 1661).

Had he graphed his table, he might have been tempted to draw a tangent and then graph that. A pity he did not, so that a statistician could have been credited with discovering derivatives. And yet, for all the things he did not do, we must give Graunt enormous credit for what he did do.

He brought, empirically, the notion of continuity into data analysis. By his tabulation of the cumulative distribution function, Graunt brought forth the modern science of statistics. No longer would stochastics be simply a plaything for the gentleman hobbyist. It would be the fundamental grammar of empirical

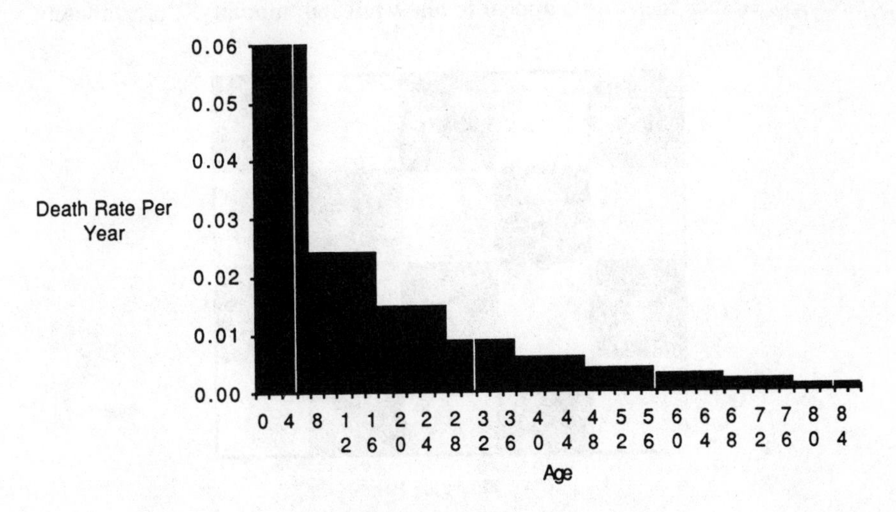

Figure 2.2.5. Graunt's histogram.

science. Graunt gave us the first rationally constructed spreadsheet. As we know from Pepys' journal and other sources, Graunt died destitute and apparently dropped from his membership (the first modern statistician had been inducted by the command of Charles II over the grumblings of other members) in the Royal Society of London.

References

Graunt, John (1662). *Natural and Political Observations on the Bills of Mortality*, London.

2.3. MODULAR WARGAMING

Checkerboard-based games are of ancient origin, being claimed by a number of ancient cultures (Figure 2.3.1). One characteristic of these games is the restricted motion of the pieces, due to the shape and tiling of the playing field. This is overcome, in measure, in chess by giving pieces varying capabilities for motion both in direction and distance. Another characteristic of these games is their essential equality of firepower. A pawn has the same power to capture a queen as the queen to capture a pawn. Effectiveness of the various pieces is completely a function of their mobility.

The directional restrictions of square tiles are a serious detriment to checkerboard games if they are to be reasonable simulations of warfare. The most satisfactory solution, at first glance, would be to use building blocks based on circles, since such tiles would appear to allow full 360° mobility. Unfortunately,

Figure 2.3.1

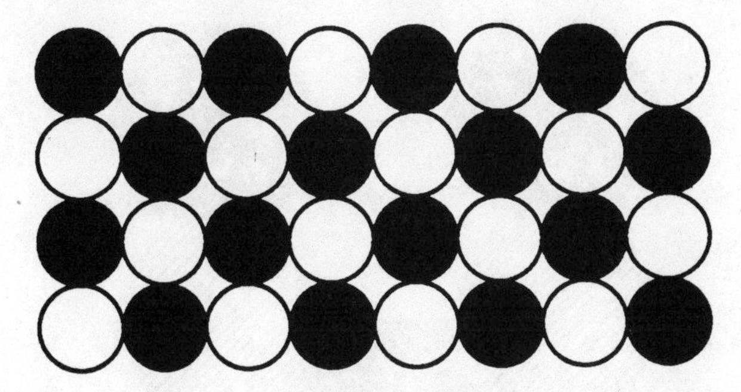

Figure 2.3.2

as we observe in Figure 2.3.2, circles cannot be satisfactory tiles, since they leave empty spaces between the tiles.

A natural first attempt to overcome the difficulty of circles as tiles would be to use equilateral octagons, since these allow motion to the eight points of the compass, N, NE, E, SE, S, SW, W, NW. Unfortunately, as we see in Figure 2.3.3, this still leaves us with the empty space phenomenon.

Figure 2.3.3

Figure 2.3.4

None of the ancient games is particularly apt as an analogue of combat after the development of the longbow, let alone after the invention of gunpowder. Accordingly, the Prussian von Reiswitz began to make suitable modifications leading in 1820 to Kriegspiel. The variants of the Prussian game took to superimposing a hexagonal grid over a map of actual terrain (see Figure 2.3.4).

Motion of various units was regulated by their capabilities in their particular terrain situation. The old notion of "turns" was retained, but at each turn, a player could move a number of units subject to a restriction on total move credits. Combat could be instituted by rules based on adjacency of opposing forces. The result of the combat was regulated by the total firepower of the units involved on both sides in the particular terrain situation. A roll of the dice followed by lookup in a combat table gave the casualty figures together with advance and retreat information.

The Prussian game, together with later American variants, such as Strategos, was validated against actual historical combat situations. In general, these games were excellent in their ability to simulate the real-world situation. Their major difficulty was one of bookkeeping. Frequently, a simulated combat could take longer to play than the actual historical battle. If the masking of movements and questions of intelligence gathering were included in the game, a large number of referees was required.

In attempting to take advantage of the computer, the creators of many modern military wargames have attempted to go far beyond resolution of the bookkeeping problems associated with Kriegspiel. Very frequently, these games do not allow for any interaction of human participants at all.

Initial conditions are loaded into a powerful mainframe computer, and the machine plays out the game to conclusion based on a complex program that may actually look at the pooled result of simulations of individual soldiers firing at each other, even though the combat is for very large units. Any real-time corrections for imperfections in the game are accordingly impossible. Any training potential of such games is obviously slight.

Furthermore, the creators of many of these games may disdain to engage in any validation based on historical combat results. Such validation as exists may be limited to checking with previous generations of the same game to see whether both gave the same answer.

If we know anything about artificial intelligence (and admittedly we know very little), it would apppear to be that those simulations work best which seem to mimic the noncomputerized human system. Attempts to make great leaps forward without evolution from noncomputerized systems are almost always unsuccessful. And it is another characteristic of such a nonevolutionary approach that it becomes quickly difficult to check the results against realistic benchmarks. Before anyone realizes it, a new, expensive, and, very likely, sterile science will have been created soaking up time and treasure and diverting us from the real-world situation.

My own view is that it is better to use the computer as a means of alleviating the bookkeeping difficulties associated with Kriegspiel-like board games. In the late 1970s and early 1980s, I assigned this task to various groups of students at Rice University. Experience showed that 200 person hours of work generally led to games that could emulate historical results very well.

At least another 500 person hours would have been required to make these games user friendly, but the rough versions of the games were instructive enough. One criticism made against historical validation is that technology is advancing so rapidly that any such validations are meaningless. It is claimed that the principal function of wargaming ought to be predictions of what will happen given the new technologies. While not agreeing that parallels between historical situations and future conflicts are irrelevant (and I note here that the strategy and tactics hobbyists generally make games ranging from Bronze

Age warfare to starship troopers), I agree that the predictive aspect, in the form of scenario analyses, is very important.

Accordingly, one student created a game for conflict between an American carrier task force and a Soviet missile cruiser task force. Given the relatively close-in combat that would be likely, it appeared that if the Soviet commander is willing to sacrifice his force for the much more costly American force, he can effect an exchange of units by a massive launch of missles at the outset of the conflict. Clearly, such a playout could have serious technological implications, for example, the desirability of constructing a system of jamming and antimissile defenses which is highly resistant to being overwhelmed by a massive strike. Or, if it is deemed that such a system could always be penetrated by further technological advances on the Soviet side, it might be appropriate to reconsider task forces based around the aircraft carrier. In any event, I personally would much prefer an interactive game in which I could see the step-by-step results of the simulation.

Also, a validation using, say, data from the Falkland conflict could be used to check modular portions of the game. World War II data could be used to check other parts. The validation would not be as thorough as one might wish, but it would be a goodly improvement on no validation at all. Some "supersophisticated" unvalidated computer simulation in which the computer simply played with itself and, at the end of the day, told me that existing antimissile defenses were sufficient would leave me neither comforted nor confident.

An integral part of any Kriegspiel computerization should deal with the resolution of the likely results of a conflict. A ready means of carrying this out was made available via the famous World War I opus of Lanchester. Let us suppose that there are two forces, the Blue and the Red, each homogeneous, and with sizes u and v, respectively.

Then, if the fire of the Red force is directed, the probability a particular Red combatant will eliminate some Blue combatant in time interval $[t, t + \Delta t]$ is given simply by

$$P(\text{Blue combatant eliminated in } [t, t + \Delta t]) = c_1 \Delta t, \tag{2.3.1}$$

where c_1 is the Red coefficient of directed fire. If we wish then to obtain the total number of Blue combatants eliminated by the entire Red side in $[t, t + \Delta t]$, we simply multiply by the number of Red combatants to obtain

$$E(\text{change in Blue in } [t, t + \Delta t]) = -vc_1 \Delta t. \tag{2.3.2}$$

Replacing u by its expectation (as we have the right to do in many cases where the coefficient is truly a constant and v and u are large), we have

$$\frac{\Delta u}{\Delta t} = -c_1 v. \tag{2.3.3}$$

This gives us immediately the differential equation

$$\frac{du}{dt} = -c_1 v. \tag{2.3.4}$$

Similarly, we have for the Red side

$$\frac{dv}{dt} = -c_2 u. \tag{2.3.5}$$

This system has the time solution

$$u(t) = u_0 \cosh \sqrt{c_1 c_2}\, t - v_0 \sqrt{\frac{c_1}{c_2}} \sinh \sqrt{c_1 c_2}\, t \tag{2.3.6}$$

$$v(t) = v_0 \cosh \sqrt{c_1 c_2}\, t - u_0 \sqrt{\frac{c_2}{c_1}} \sinh \sqrt{c_1 c_2}\, t.$$

A more common representation of the solution is obtained by dividing (2.3.4) by (2.3.5) to obtain

$$\frac{du}{dv} = \frac{c_1 v}{c_2 u}, \tag{2.3.7}$$

with solution

$$u^2 - u_0^2 = \frac{c_1}{c_2}(v^2 - v_0^2). \tag{2.3.8}$$

Now u and v are at "combat parity" with each other when

$$u^2 = \frac{c_1}{c_2} v^2. \tag{2.3.9}$$

(A special point needs to be made here. Such parity models assume that both sides are willing to bear the same proportion of losses. If such is not the case, then an otherwise less numberous and less effective force can still emerge victorious. For example, suppose that the Blue force versus Red force coefficient is 0.5 and the Blue force has only 0.9 the numerosity of the Red force.

Then if Blue is willing to fight until reduced to 0.2 of his original strength, but Red will fight only to 0.8 of his original strength, then using (2.3.8) we find that by the time Red has reached maximal acceptable losses, Blue still has 30% of his forces and thus wins the conflict. This advantage to one force to accept higher attrition than his opponent is frequently overlooked in wargame analysis. The empirical realization of this fact has not escaped the attention of guerrilla leaders from the Maccabees to the Mujaheddin.)

Accordingly, it is interesting to note that if there is a doubling of numbers on the Red side, Blue can only maintain parity by seeing to it that c_2/c_1 is quadrupled, a seemingly impossible task.

Lanchester's formula for undirected fire follows from similar Poissonian arguments. The probability that a Red combatant will eliminate some Blue combatant in $[t, t + \Delta t]$ is given by

$$\begin{aligned} (2.3.10) \qquad & P(\text{a Blue eliminated by a Red in } [t, t + \Delta t]) \\ & = P(\text{shot fired in } [t, t + \Delta t])P(\text{shot hits a Blue})\Delta t. \end{aligned}$$

Now, the probability that a shot aimed at an area rather than an individual hits someone is proportional to the density of Blue combatants in the area, and hence proportional to u. Thus, we have

$$(2.3.11) \qquad P(\text{Blue eliminated in } [t, t + \Delta t]) = d_1 u \, \Delta t.$$

The expected number of Blues eliminated in the interval is given by multiplying the above by the size of the Red force, namely, v. So the differential equations are

$$(2.3.12) \qquad \frac{du}{dt} = -d_1 uv \quad \text{and} \quad \frac{dv}{dt} = -d_2 uv.$$

This system has the time solution

$$(2.3.13a) \qquad u(t) = \frac{(d_2/d_1)u_0 - v_0}{(d_2/d_1) - (v_0/u_0)e^{-(d_2 u_0 - d_1 v_0)t}}$$

and

$$(2.3.13b) \qquad v(t) = \frac{(d_1/d_2)v_0 - u_0}{(d_1/d_2) - (u_0/v_0)e^{-(d_1 v_0 - d_2 u_0)t}}.$$

Here, when dividing the equations in (2.3.12) and solving, we obtain the parity equation:

$$u - u_0 = \frac{d_1}{d_2}(v - v_0). \tag{2.3.14}$$

In such a case, a doubling of Red's parity force can be matched by Blue's doubling of d_2/d_1.

In attempting to match either law (or some other) against historical data, one needs to be a bit careful. In 1954, Engel claimed to have validated the applicability of Lanchester's directed fire law for the Battle of Iwo Jima. He used no records for Japanese casualties and simply adjusted the two parameters to fit the record of American casualty data. According to Engel's model, Japanese resistance was completely eliminated in 36 days. But American data reveal that resistance continued for some time after the battle was over, with 20 Japanese killed in a single banzai charge on day 37 and up to 70 prisoners taken on some days up to 1 month after day 36. More significantly, there is available partial Japanese force data delivered by the Japanese commander, General Kuribayashi, by radio to Tokyo at times well before day 36. For example, on day 21 of the conflict, when Engel's model gives a figure of 8550 for the Japanese forces on the island, Kuribayashi gives the actual figure of 1500. Using the partial Japanese casualty records shows that the directed fire model gave answers much at variance with the data (sometimes off the Japanese total effectives by a factor of 4) and that the undirected fire model appeared to work much more satisfactorily. It is possible to track very closely the American force levels using either the directed or undirected fire models. But the undirected fire model has the additional attribute of closely tracking the partial force information for the Japanese. We have exhibited both the directed and undirected fire models above in Figure 2.3.5.

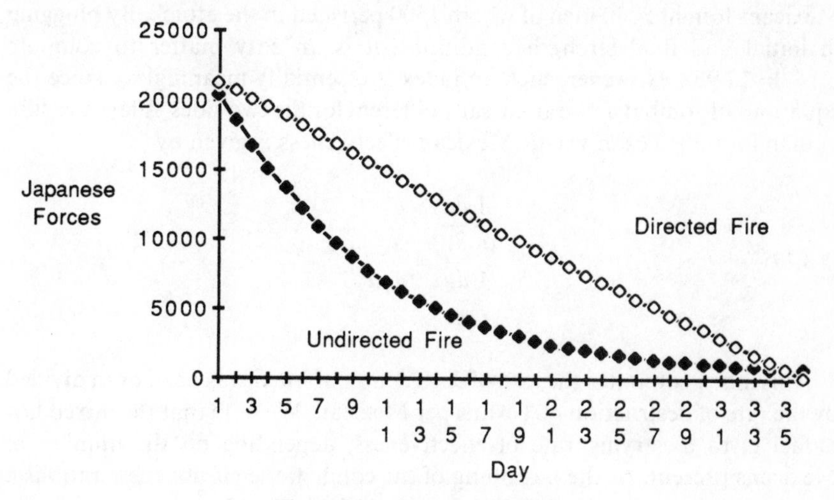

Figure 2.3.5. Battle of Iwo Jima.

In the above, it is rather clear that the undirected fire model is to be preferred over the directed one. However, any homogeneous force model would probably not be as satisfactory as a heterogeneous force model in an engagement in which naval gunfire together with marine assault both played important roles. We shall address the heterogeneous force model problem shortly. In a much broader context of combat simulation, we note that a model which appears at first glance to do an excellent job of "prediction" may become seriously deficient as more data are made available.

The year 1986 marked the 150th anniversary of the Battle of the Alamo. This battle gives an example of a situation in which a mixture of the two models is appropriate. Since the Texans were aiming at a multiplicity of Mexican targets and using rifles capable of accuracy at long range (300 m), it might be appropriate to use the directed fire model for Mexican casualties. Since the Mexicans were using less accurate muskets (100 m) and firing against a fortified enemy, it might be appropriate to use the undirected fir model for Texan casualties. This would give

$$\frac{du}{dt} = -d_1 uv \quad \text{and} \quad \frac{dv}{dt} = -c_2 u. \tag{2.3.15}$$

The parity equation is given by

$$v^2 - v_0^2 = \frac{2c_2}{d_1}(u - u_0). \tag{2.3.16}$$

The Texans fought 188 men, all of whom perished in the defense. The Mexicans fought 3000 men of whom 1500 perished in the attack. By plugging in initial and final strength conditions, it is an easy matter to compute $c_2/d_1 = 17{,}952$. However, such an index is essentially meaningless, since the equations of combat are dramatically different for the two sides. A fair measure of man for man Texan versus Mexican effectiveness is given by

$$\frac{\dfrac{1}{u}\dfrac{dv}{dt}}{\dfrac{1}{v}\dfrac{du}{dt}} = \frac{c_2}{d_1 u}. \tag{2.3.17}$$

This index computes the rate of destruction of Mexicans per Texan divided by the rate of destruction of Texans per Mexican. We note that the mixed law model gives a varying rate of effectiveness, depending on the number of Mexicans present. At the beginning of the conflict, the effectiveness ratio is a possible 96; at the end, a romantic but unrealistic 17,952.

The examination of this model in the light of historical data should cause us to question it. What is wrong? Most of the Mexican casualties occurred before the walls were breached. Most of the Texan casualties occurred after the walls were breached. But after the walls were breached, the Mexicans would be using directed fire against the Texans.

We have no precise data to verify such an assumption, but for the sake of argument, let us assume that the Texans had 100 men when the walls were breached, the Mexicans 1800. Then (2.3.16) gives $c_2/d_1 = 32{,}727$. The combat effectiveness ratio $c_2/d_1 u$ goes then from 174 at the beginning of the siege to 327 at the time the walls were breached. For the balance of the conflict we must use Eqs. (2.3.4) and (2.3.5) with the combat effectiveness ratio $c_2/c_1 = 99$ (computed from Eq. (2.3.8)). Personally, I am not uncomfortable with these figures. The defenses seem to have given the Texans a marginal advantage of around 3. Those who consider the figures too "John Wayneish" should remember that the Mexicans had great difficulty in focusing their forces against the Alamo, whereas the Texans were essentially all gainfully employed in the business of fighting. This advantage to a group of determined Palikari to defend a fortified position against overwhelming numbers of a besieging enemy is something we shall return to shortly.

Another example of the effect of fortifications in combat is obtained from the British–Zulu War of 1879. On January 22, at Isandhlwana, an encamped British column of 1800 British soldiers and 1000 native levies was attacked by 10,000 warriors of the Zulu King Cetawayo. The suggestion of the Afrikaaner scouts to laager (roughly drawing the wagons into a circle) was rejected by the British commander. Consequently, even though the British troops had the benefit of modern breech-loading rifles, they were quickly engaged in hand-to-hand combat by the Zulus. The result of the conflict was that only around 55 British and 300 of the native levies survived. We do not have precise knowledge of the Zulu losses; but we do know that, on the evening of January 22, 5000 of the same Zulu force attacked a British force of 85 at Rorkesdrift. The British commander (a lieutenant of engineers, John Chard, later a major general) had used the few hours warning to laager his camp with overturned wagons and sacks of meal. On January 23, the Zulus withdrew, leaving 400 dead on the field. British losses were 17 killed and 10 seriously wounded. Here we have an example of nearly identical types of forces on the attack and on the defense in both engagements. Since the Zulus always fought hand-to-hand, we shall use (2.3.4) and (2.3.5) in both battles for both sides. If we assume (a popular notion of the day) that the native levies made no contribution, and that 5000 Zulus were incapacitated by the Isandhlwana engagement, the combat effectiveness of British soldier versus that of Zulu soldier computes to be 23.17. (The assumptions here obviously tend hugely to inflate the actual British versus Zulu combat effectiveness.) At Rorkesdrift, the combat effectiveness ratio goes to 994.56. Thus, the advantage given to the British defenders of Rorkesdrift by the hastily constructed defenses was at least 994.56/23.17 = 42.92. The advantage was not primarily an increased combat effectiveness of the British soldiers, but rather a diminution of the combat effectiveness of the Zulus. Having transmitted some feeling as to the advantages of commonsense utilization of the method of Lanchester (borrowed in spirit from Poisson), we shall now take the next step in its explication: namely, the utilization of heterogeneous force equations.

Let us suppose that the Blue side has m subforces $\{u_j\}; j = 1, 2, \ldots, m$. These might represent artillery, infantry, armor, and soon. Also, let us suppose that the Red side has n subforces $\{v_i\}$; $i = 1, 2, \ldots, n$. Then the directed fire equations (2.3.4) and (2.3.5) become

$$\frac{du_j}{dt} = -\sum_{i=1}^{n} k_{ij} c_{1ij} v_i \tag{2.3.18}$$

and

$$\frac{dv_i}{dt} = -\sum_{j=1}^{m} l_{ji} c_{2ji} u_j. \tag{2.3.19}$$

Here, k_{ij} represents the allocation (a number between 0 and 1 such that $\sum_{j=1}^{m} k_{ij} \leq 1$) of the ith Red subforce's firepower against the jth Blue subforce. c_{1ij} represents the Lanchester attrition coefficient of the ith Red subforce against the jth Blue subforce. Similar obvious definitions hold for $\{l_{ji}\}$ and $\{c_{2ji}\}$.

Equation (2.3.18) furnishes us a useful alternative to the old table lookup in Kriegspiel. Numerical integration enables us to deal handily and easily with any difficulties associated with turn to turn changes in allocation and effectiveness, reinforcements, and so on. Experience has shown that computerized utilization of mobility rules based on hexagonal tiling superimposed on actual terrain, together with the use of (possibly randomized) Lanchester heterogeneous force combat equations, makes possible the construction of realistic war games at modest cost.

Beyond the very real utility of the Lanchester combat laws to describe the combat mode for war games, they can be used as a model framework to gain insights as to the wisdom or lack thereof of proposed changes in defense policy. For example, a dismantling of intermediate range missiles in eastern and central Europe throws additional responsibility on the effectiveness of NATO conventional forces, since a conventional Soviet attack is no longer confronted with a high risk of a Pershing missile attack from West Germany against Russia. The rather large disparity in conventional forces between the Soviet block and NATO forces can roughly be addressed by a consideration of Lanchester's directed fire model. As we have observed in (2.3.9), in the face of a twofold personnel increase of Red beyond the parity level, Blue can, assuming Lanchester's directed fire model, maintain parity only by quadrupling c_2/c_1. This has usually been perceived to imply that NATO must rely on its superior technology to match the Soviet threat by keeping c_2 always much bigger than c_1.

Since there exists evidence to suggest that such technological superiority does not exist at the conventional level, it appears that the Soviets keep out of western Europe because of a fear that a conventional juggernaut across western Europe would be met by a tactical nuclear response from West Germany, possibly followed, *in extremis,* by a strategic attack against population centers in Russia: thus, the big push by the Soviets and their surrogates for "non first use of nuclear weapons" treaties and their enthusiastic acceptance of American initiatives to remove intermediate range missiles from eastern and central Europe. It is not at all unlikely that the Soviets could take western Europe in a conventional war in the absence of intermediate range missiles in West Germany if the current disparate numerical advantage of conventional Soviet forces in Europe were maintained. Thus, one practical consequence for NATO of the dismantling of intermediate range missiles in Europe might be attempts by the Western powers to bring their conventional forces to numerical parity with those of the Soviets. This might require politically difficult policy changes in some NATO countries, such as reinstitution of the draft in the United States.

In my paper "An Argument for Fortified Defense in Western Europe," I attempted to show how the c_2/c_1 ratio could be increased by using fortifications to decrease c_1. Whether or not the reader judges such a strategy to be patently absurd, it is instructive to go through the argument as a means of explicating the power of Lanchester's laws in scenario analysis.

My investigation was motivated, in part, by the defense of the Westerplatte peninsula in Gdańsk by 188 Polish soldiers from September 1 through September 7 in 1939, and some interesting parallels with the much lower tech siege of the Alamo 100 years earlier. (Coincidentally, the number of Polish defenders was the same as the number of Texans at the Alamo.) The attacking German forces included a battalion of SS, a battalion of engineers, a company of marines, a construction battalion, a company of coastal troops, assorted police units, 25 Stukas, the artillery of the Battleship Schleswig-Holstein, eight 150 mm howitzers, four 210 mm heavy mortars, 100 machine guns, and two trainloads of gasoline (the Germans tried to flood the bunkers with burning gasoline).

The total number of German troops engaged in combat during the 7 day seige was well over 3000. Anyone who has visited Westerplatte (as I have) is amazed with the lack of natural defenses. It looks like a nice place for a walkover. It was not.

The garrison was defended on the first day by a steel fence (which the Germans and the League of Nations had allowed, accepting the excuse of the Polish commander, Major Sucharski, that the fence was necessary to keep the livestock of the garrison from wandering into Gdańsk), which was quickly obliterated. Mainly, however, the structural defenses consisted in concrete fortifications constructed at the ground level and below. Theoretically, the structural fortifications did not exist, since they were prohibited by the League of Nations and the peninsula was regularly inspected by the Germans to ensure compliance. However, extensive "coal and storage cellars" were permitted, and it was such that comprised the fortifications. The most essential part of the defenses was the contingent of men there. Unlike the Texans at the Alamo who realized they were going to die only after reinforcements from Goliad failed to arrive and the decision was made not to break through Santa Anna's encirclement, the Polish defenders of Westerplatte realized, long before the conflict, that when the German invasion began, they would be doomed. It is interesting to note the keen competition that existed to gain the supreme honor of a posting to Westerplatte. Perhaps "no bastard ever won a war by dying for his country" but the defenders of the Alamo and those of Westerplatte consciously chose their deaths as an acceptable price for wreaking a bloody vengeance on the enemies of their people.

Ever since the abysmal failure of the Maginot Line in 1940, it has been taken for granted that any strategy based on even the partial use of fixed defenses is absurd. I question this view. Historically, fixed defenses have proved more effective as islands rather than as flankable dikes. The Maginot Line was clearly designed as a dike, as was the Great Wall of China, and both proved failures. It is unfortunate that the dikelike tactics of trench warfare had proved so effective in World War I. Otherwise, the French would undoubtedly have noted that they were basing their 1940 defense on a historically fragile strategy. Dikes generally can withstand force only from the front, as the Persians (finally) discovered at Thermopolae. If the dikes are sufficiently narrow and thick, however, they may be effective islands and very difficult to outflank. It was conceded by the panzer innovator, von Manstein, that Germany absolutely could not have taken the Sudentenland defenses in 1938 had they been used. This brings up another interesting point. An effective system of fixed defenses is very much dependent on the will of the people using them.

Historical examples, modern as well as ancient, of successful use of constructed defensive positions can be given ad infinitum. Among the crusading orders, the Templars and Hospitalers early discovered that they could maintain an effective Christian presence in the Holy Land only by concentrating a large percentage of their forces in a number of strongly fortified castles. This gave them sufficient nuisance value to cause concessions by the Muslim leaders. Most of the military disasters to the orders were the result of their frequent willingness to strip their castle defenses and join the crusader barons in massive land battles against numerically overwhelming odds—as at Hattin. For over 1000 years, some of the Christian peoples in the Near East, for example, the Armenians and the Maronites, maintained their very identity by mountain fortifications.

It is interesting to note that one of the crusader fortresses—Malta—never fell to the Muslims and was only taken (by treachery) by Napoleon in 1798. In the second World War, the connection between the resistance of Malta and the ultimate destruction of the Afrika Korps is well remembered. Even light, hastily constructed defenses, manned by people who do not know they are supposed to surrender when surrounded, can be extremely effective in slowing down the enemy advance, as proved by the 101st Airborne during the Battle of the Bulge. In the examples above, there seem to be some common points. First, fortified defense gives a ready means of increasing the ratio of the Lanchester coefficients in favor of the Blue side. One natural advantage to this type of defense is the fact that the defender can increase his Lanchester attrition ratio by a policy of construction over a period of time. This may be a more fruitful policy than placing all one's hopes on increasing ones Lanchester ratio by the design of new weapons systems.

Second, fortified defense should rely on adequate stores of supplies located within the "fortress perimeter." It should be assumed by the defenders that they will be completely surrounded by the enemy for long periods of time. (In their fortress at Magdeburg, the Teutonic Knights always kept 10 years' provisions for men and horses.)

Third, fortified defense is a task best undertaken by well-trained professionals with strong group loyalty.

Fourth, fortified defense is most effective when there are allied armies poised to strike the enemy at some future time and place. The fortress and the mobile striking force complement each other in their functions. The function of the fortress is to punish, harass, and divide the enemy and to maintain a presence in a particular area. In general, however, offensive activities must be left to the mobile forces. The deployment of enemy forces to take fortified positions will weaken their ability to withstand mobile offensive operations. Let us now examine modified versions of (2.3.4) and (2.3.5):

$$\frac{du}{dt} = -c_1^* v \tag{2.3.20}$$

and

$$\frac{dv}{dt} = -c_2^* u. \tag{2.3.21}$$

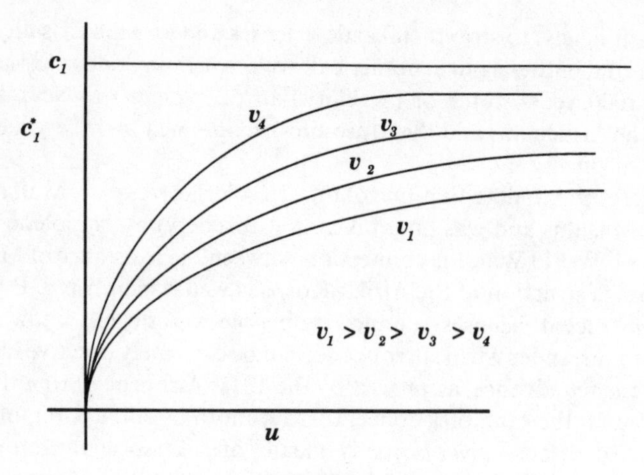

Figure 2.3.6

The attrition to Blue coefficient is represented by the variable $c_1^* = c_1^*(u, v)$ and is demonstrated graphically in Figure 2.3.6.

In the above, we assume that c_1^* never exceeds c_1, the attrition constant corresponding to nonfortified combat. Clearly, the functions c_1^* and c_2^* are functions of the manner in which the fortress has been constructed. It may be desirable to design the fortifications so that c_1^* is small, even at the expense of decreasing c_2^*. Generally, one might assume that c_2^* is close to the nonfortified attrition rate of u against v, since the defenders will have removed potential cover for the Red side. In fortress defense, the solution in time is likely to be important, since a primary objective is to maintain a Blue presence for as long as possible. Next, consider a linear approximation to the v-level curves of $c_1^*(u, v)$ (Figure 2.3.7). Then we would have

$$\frac{du}{dt} = -g(v)uv - c_1^{**}u, \tag{2.3.22}$$

where $c_1^*(u, v) = g(v)u$ and c_1^{**} is the Blue coefficient of internal attrition. (We note that this analysis has moved us, quite naturally, to an undirected fire model for the defenders' losses. The model thus derived is essentially that used earlier for the Alamo.) We might reasonably expect that the besieging forces would maintain more or less a constant number of troops in the vicinity of the redoubt. Hence, we would expect

$$\frac{dv}{dt} = -c_2^*u - c_2^{**}v + P(u, v) = 0, \tag{2.3.23}$$

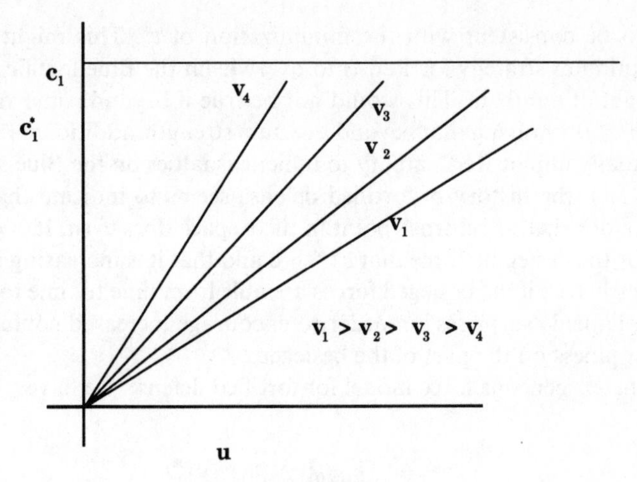

Figure 2.3.7

where $P(u, v)$ is the rate of replacement necessary to maintain constant v strength and c_2^{**} is the Red coëfficient of internal attrition. We might expect that $c_2^{**} \gg c_1^{**}$, since inadvertent self-inflicted casualties are a well-known problem for the besieging force. Then

$$u(t) = u_0 \exp[-(g(v)v + c_1^{**})t]. \tag{2.3.24}$$

The enemy attrition by time t is given by

$$\int_0^t P(u, v)\,dt = c_2^{**}tv + c_2^* u_0 \frac{1 - \exp[-(g(v)v + c_1^{**})t]}{g(v)v + c_1^{**}}. \tag{2.3.25}$$

If the Blue defense can hold out until $u = \gamma u_0$ (where $0 < \gamma < 1$), then the time until the end of resistance is given by

$$t^* = -\frac{\ln(\gamma)}{g(v)v + c_1^{**}}. \tag{2.3.26}$$

We have then that the total losses to the Red side by the time the defense falls are given by

$$\frac{c_2^* u_0 (1 - \gamma) - c_2^{**} v \ln(\gamma)}{g(v)v + c_1^{**}}. \tag{2.3.27}$$

It is interesting to note that if $c_2^{**} = 0$, then the minimization of Red casualties appears to be consistent with the minimization of t^*. This might indicate that an optimum strategy for Red is to overwhelm the Blue fortifications by shear weight of numbers. This would not be true if beyond some value of v, $d(g(v)v)/dv \le 0$, implying that beyond a certain strength, additional Red forces would actually impair Red's ability to inflict casualties on the Blue side. As a matter of fact, the history of fortified defense seems to indicate that such a "beginning of negative returns" point in the v space does exist. It is generally the case for the besieging force that $c_2^{**} \gg 0$ and that it is increasing in v. This is particularly true if the besieged forces are able from time to time to conduct carefully planned "surprises" in order to encourage increased confusion and trigger happiness on the part of the besiegers.

In the heterogeneous force model for fortified defense, we have

$$\frac{du_j}{dt} = -\sum_{i=1}^{n} k_{ij} g_{ij}(v_i) v_i u_j - c_{1j}^{**} u_j \tag{2.3.28}$$

and

$$\frac{dv_i}{dt} = -\sum_{j=1}^{m} l_{ji} c_{2ji}^{*} u_j - c_{2i}^{**} v_i. \tag{2.3.29}$$

The size of the jth Blue subforce at time t is given by

$$u_j(t) = u_j(0) \exp\left(-t \sum_{i=1}^{n} k_{ij} g_{ij}(v_i) v_i + c_{ij}^{**} \right). \tag{2.3.30}$$

The total attrition to the ith enemy subforce at time t is given by

$$\begin{aligned} \int_0^t P_i(u, v)\, d\tau &= \sum_{j=1}^{m} l_{ji} c_{2ji}^{*} u_j(0) \int_0^t \exp\left(-\tau \sum_{i=1}^{n} k_{ij} g_{ij}(v_i) v_i + c_{ij}^{**} \right) d\tau + c_{21}^{**} t v_i \\ &= \sum_{j=1}^{m} l_{ji} c_{2ji}^{*} u_j(0) \frac{1 - \exp\left(-t \sum_{i=1}^{n} k_{ij} g_{ij}(v_i) v_i \right)}{\sum_{i=1}^{n} k_{ij} g_{ij}(v_i) v_i + c_{1j}^{**}} + c_{2i}^{**} t v_i. \end{aligned} \tag{2.3.31}$$

Suppose that the effectiveness (at time t) of the Blue defender is measured by

$$T(t) = \sum_{j=1}^{m} a_j u_j(t), \tag{2.3.32}$$

where the a_j are predetermined relative effectiveness constants. If we assume that the fortress is lost when the effectiveness is reduced to some fraction γ of its initial value, that is, when

$$T(t) < \gamma T(0), \tag{2.3.33}$$

then we can use (2.3.31), in straightforward fashion, to solve for the time of capture.

The above model gives some indication of the power of the simple Lanchester "laws" in analyzing a "what if" scenario. It is, in large measure, the lack of "gee-whizziness" of Lanchester's models which renders them such a useful device to the applied worker. Generally, after a few hours of self-instruction, potential users can reach the level of sophistication where they can flowchart their own wargames or other forms of scenario analysis.

References

Dupuy, R. Ernest and Dupuy, Trevor N. (1970). *The Encyclopedia of Military History,* Harper & Row, New York.

Engel, J. H. (1954). A verification of Lanchester's law, *Operations Research,* **12,** 344–358.

Lanchester, F. W. (1916). Mathematics in warfare, reprinted in *The World of Mathematics,* Vol. **4,** J. R. Newman, ed., Simon and Schuster, New York, 1956, pp. 2138–2157.

Newcomb, Richard F. (1965). *Iwo Jima,* Holt, Rinehart and Winston, New York.

Thompson, J. R. (1979). An example of data-poor model validation, in *Decision Information,* C. P. Tsokos and R. M. Thrall, eds., Academic, New York, pp. 405–408.

Thompson, J. R. (1979). An argument for fortified defense in western Europe, in *Decision Information,* C. P. Tsokos and R. M. Thrall, eds., Academic, New York, pp. 395–404.

2.4. PREDATION AND IMMUNE RESPONSE SYSTEMS

Let us consider Volterra's predator–prey model and some consequences for modeling the human body's anticancer immune response system. For the classical shark–fish model, we follow essentially Haberman (1977). Suppose we have predators, say sharks, whose numbers are indicated by S, who prey on, say fish, whose numbers are indicated by F. In the 1920s, it was brought to the attention of Volterra that there appeared to be a periodic pattern in the abundance of certain food fish in the Adriatic, and that this pattern did not appear to be simply seasonal. Volterra attempted to come up with the simplest logical explanation of this periodicity.

We might suppose that the probability a typical shark gives birth to another shark (for reasons of simplicity we treat the sharks as though they were single-cell creatures) is given by

$$P(\text{birth in } [t, t + \Delta t]) = (\lambda F)\Delta t. \tag{2.4.1}$$

Here the assumption is that the probability of reproduction is proportional to the food supply, that is, to the size of the fish population. The probability a shark dies in the time interval is considered to be a constant $k\,\Delta t$. Thus, the expected change in the predator population during $[t, t + \Delta t]$ is given by

$$E[\Delta S] = S(\lambda F - k)\Delta t. \tag{2.4.2}$$

As we have in the past, we shall assume that for a sufficiently large predator population, we may treat the expectation as essentially deterministic. This gives us the differential equation

$$\frac{dS}{dt} = S(\lambda F - k). \tag{2.4.3}$$

Similarly, the probability that a given fish will reproduce in $[t, t + \Delta t]$ minus the probability it will die from natural causes may be treated like

$$P(\text{birth in } [t, t + \Delta t]) = a\,\Delta t. \tag{2.4.4}$$

We have assumed that the fish have essentially an unlimited food supply. The death by predation, on a per fish basis, is obviously the number of sharks multiplied by their fish eating rate, c, giving the differential equation

$$\frac{dF}{dt} = F(a - cS). \tag{2.4.5}$$

Now the system of equations given by (2.4.3) and (2.4.5) has no known simple time domain solution, although numerical solution is obviously trivial. However, let us examine the F versus S situation by dividing (2.4.5) by (2.4.3). This gives us

$$\frac{dF}{dS} = \frac{F}{\lambda F - k}\,\frac{a - cS}{S}. \tag{2.4.6}$$

The solution to (2.4.6) is easily seen to be

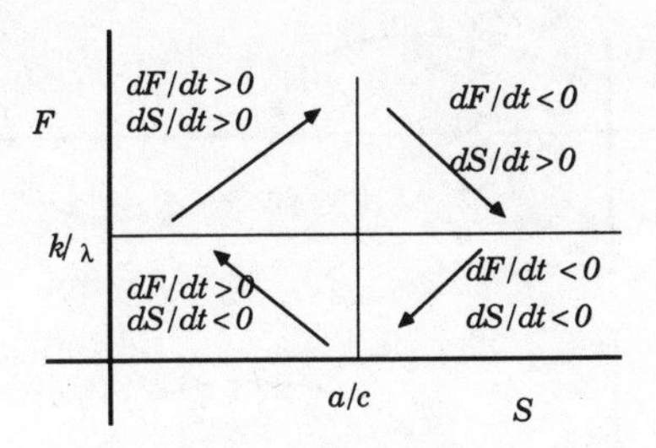

Figure 2.4.1

(2.4.7) $$F^{-k}e^{\lambda F} = Ee^{-cS}S^{a},$$

with E a constant. Now, let us use (2.4.3) and (2.4.5) to trace the path of F versus S. We note first that $F = k/\lambda$ gives an unchanging S population; $S = a/c$ gives an unchanging F population.

The consequences of Figure 2.4.1 are that the F versus S plot must either be a closed repeating curve or a spiral. We can use (2.4.7) to eliminate the possibility of a spiral. Let us examine the level curves of F and S corresponding to the common Z values in

(2.4.8) $$F^{-k}e^{\lambda F} = Ee^{-cS}S^{a} = Z.$$

In Figure 2.4.2, we sketch the shapes of Z versus F and S, respectively, and

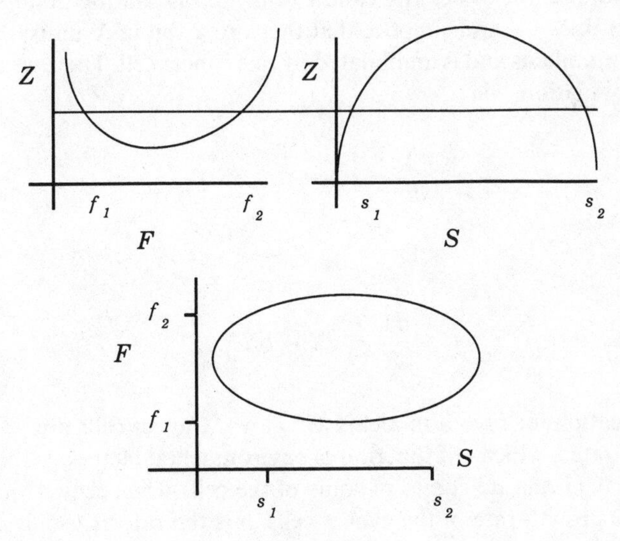

Figure 2.4.2

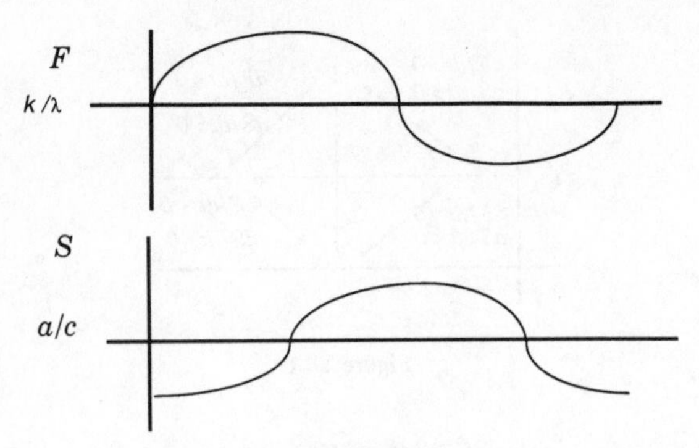

Figure 2.4.3

use these values to trace the F versus S curve. We note that since each value of Z corresponds to at most four points on the F versus S curve, a spiral structure is out of the question, so we obtain the kind of closed curve that was consistent with the rough data presented to Volterra. Using Figure 2.4.1 in conjunction with Figure 2.4.2, we can sketch the time behavior of the two populations (Figure 2.4.3). Here we note periodic behavior with the fish curve leading the shark curve by 90°.

Let us now turn to an apparently quite different problem, that of modeling the body's immune response to cancer. Calling the number of cancer cells X, let us postulate the existence of antibodies in the human organism which identify and attempt to destroy cancer cells. Let us call the number of these "immunoentities" Y, and suppose that they are given in X units; that is, one unit of Y annihilates and is annihilated by one cancer cell. Then, we can model the two populations via

$$\frac{dX}{dt} = \lambda + aX - bXY \tag{2.4.9}$$

and

$$\frac{dY}{dt} = cX - bXY. \tag{2.4.10}$$

The justification for such a model is as follows. Cancer cells are produced at a constant rate λ which is a function of environmental factors, inability of the body to make accurate copies of some of the cells when they divide, and so on. a is the growth rate of the cancer cells. b is the rate at which antibodies attack and destroy the cancer cells. c is the rate of response of the antibody population to the presence of cancer cells.

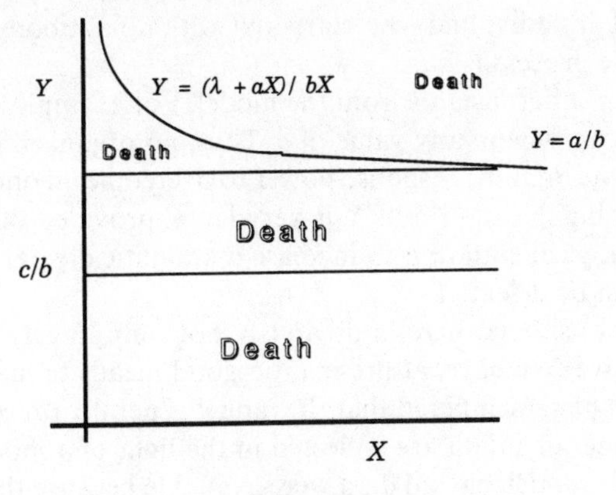

Figure 2.4.4

Although we cannot obtain closed form solutions for the system given by (2.4.9) and (2.4.10), we can sketch a system of curves that will give us some feel as to which individuals will have immune systems that can cope with the oncogenesis process. From (2.4.10), we note that Y decreases if $dY/dt = cX - bXY < 0$; that is, if $Y > c/b$. If the inequality is reversed, then Y will increase. Similarly, from (2.4.9), we note that X decreases if $dX/dt = \lambda + aX - bXY < 0$; that is, if $Y > (\lambda + aX)/bX$. Let us examine the consequences of these facts by looking at Figure 2.4.4. The prognosis here would appear to be very bad. The body is not able to fight back the cancer cells and must be overwhelmed.

On the other hand, let us examine the more hopeful scenario in Figure 2.4.5. We note the change if c increases dramatically relative to a. We now have regions where the body will arrive at a stable equilibrium of cancer cells and antibodies. We should also note that in both Figure 2.4.4 and Figure 2.4.5

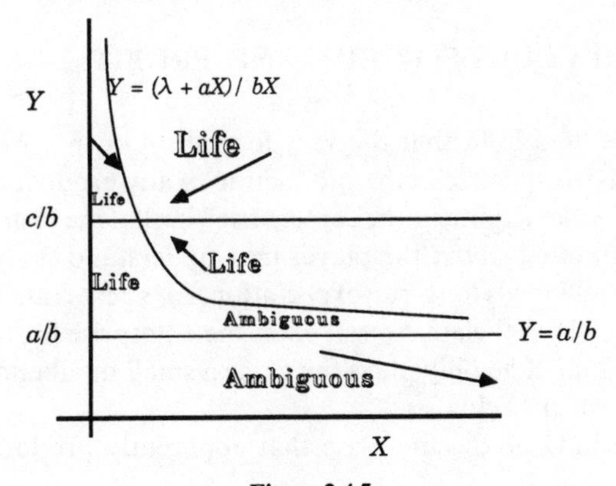

Figure 2.4.5

the situation of an individual who starts out with no antibody backup at the beginning of the process is bad.

We can glean other insights from the model. For example, a large enough value of λ can overwhelm any value of c. Thus, no organism can reasonably expect to have the immune response power to overcome all oncogenic shocks, no matter how big. Next, even if X is very large, provided only that we can change the biological situation to increase c dramatically, while suppressing λ, the tumor can be defeated.

The model considered here is obviously not only hugely simplified, but purely speculative. We have, at present, no good means of measuring X and Y. But it should be remembered that the model generally precedes the collection of data: generally, data are collected in the light of a model. In the case of Volterra's fish model, partial data were available because the selling of fish was measured for economic reasons. Volterra was, in short, fortunate that he could proceed from a well-developed data set to an explanatory model. This was serendipitous and unusual.

Generally, we waste much if we insist on dealing only with existing data sets and refuse to conjecture on the basis of what may be only anecdotal information. If we are being sufficiently bold, then for every conjecture that subsequently becomes substantiated we should expect to be wrong a dozen times. Model building is not so much the safe and cozy codification of what we are confident about as it is a means of orderly speculation.

References

Haberman, Richard (1977). *Mathematical Models,* Prentice Hall, Englewood Cliffs, NJ.

2.5. PYRAMID CLUBS FOR FUN AND PROFIT

There are those who hold that the very formalism of the "free market" will produce good—irrespective of the production of any product or service other than the right to participate in the "enterprise" itself. One example of such an enterprise is gambling. Here, the player may understand that he is engaging in an activity in which his long-run expectations for success are dim—the odds are against him. Nevertheless, he will enter the enterprise for fun, excitement, and the chance that, if he only plays the game a small number of times, he will get lucky and beat the odds.

Another example of an enterprise that apparently produces no good or service is that of the pyramid club. Unlike gambling, the pyramid club gives the participants the notion that they almost certainly will "win"; that is, their gain will exceed, by a very significant margin, the cost of their participation. Let us consider a typical club structure. For the cost of \$2000, the member is allowed to recruit up to six new members. For each member he recruits, he receives a commission of \$1000. Furthermore, each of the new members is inducted with the same conditions as those of the member who inducted them. Now for each recruit made by second-level members, the first-level member receives a commission of \$100. This member is allowed to share in these \$100 commissions down through the fifth level. Generally, there is some time limit as to how long the member has to recruit his second-level members—typically a year. Thus, his anticipated return is

$$\text{anticipated return} = 1000 \times 6 + (6^2 + 6^3 + 6^4 + 6^5) \times 100 = 938{,}400. \tag{2.5.1}$$

It is this apparent certainty of gain which attracts many to pyramid enterprises. Many state governments claim that this hope of gain is hugely unrealistic, and thus that pyramid enterprises constitute fraud. We wish to examine this claim.

Let us suppose we consider only those members of society who would become members if asked. Let us say that at any given time those who are already members will be included in the pool Y and those who have not yet joined but would if asked are included in the pool X. If we examine the probability that a member will effect a recruitment in time interval Δt, this appears to be given by

$$P(\text{recruitment in } [t, t + \Delta t]) = \frac{kX}{X + Y}\Delta t. \tag{2.5.2}$$

Here, k is the yearly rate of recruitment if all persons in the pool were nonmembers (e.g., $k = 6$). Then we have that the expected number of recruits by all members in $[t, t + \Delta t]$ is given by

$$E(\text{number of recruits in } [t, t + \Delta t]) = \frac{kXY}{X + Y}\Delta t. \tag{2.5.3}$$

We will neglect any exodus from the pool. Also, we neglect entries into the pool. Thus, if we replace the expectation of Y by Y itself, and divide by Δt, and let Δt go to 0, we have

$$\frac{dY}{dt} = \frac{kXY}{X+Y}. \tag{2.5.4}$$

Let us make the assumption that $X + Y = c$, a constant. Then we have the easily solvable (using partial fractions) equation

$$\frac{dY}{Y(c-Y)} = \frac{k}{c}dt. \tag{2.5.5}$$

So we have

$$t = \frac{1}{k}\ln\left(\frac{Y}{c-Y}\right) - \frac{1}{k}\ln\left(\frac{Y_0}{c-Y_0}\right). \tag{2.5.6}$$

Now, when $dY/dt = 0$, there is no further increase of Y. Thus, the equilibrium (and maximum) value of Y is given by

$$Y_e = c. \tag{2.5.7}$$

For the present example, the maximum value of Y, Y_e, will only be reached at $t = \infty$. But it is relevant to ask how long it will take before Y equals, say, $.99c$. If we assume that Y_0 equals $.0001c$, a little computation shows that t (when $Y = .99c$) $= 2.3$ years.

Now, the rate of recruitment per member per year at any given time is given by

$$\frac{dY/dt}{Y} = \frac{k(c-Y)}{c}. \tag{2.5.8}$$

At time $t = 2.3$, and thereafter,

$$\frac{dY/dt}{Y} \leq 0.06, \qquad \frac{dY/dt}{Y} \leq 0.06. \tag{2.5.9}$$

Unfortunately, a member who joins at $t = 1.87$ or thereafter must replace the 6 in (2.5.1) by a number no greater than .06. Thus, the anticipated return to a member entering at this time is rather less than 938,400:

(2.5.10)

$$\text{anticipated return} \leq 1000 \times .06 + (.06^2 + .06^3 + .06^4 + .06^5) \times 100 = 60.38.$$

The difference between a pyramid structure and a bona fide franchising enterprise is clear. In franchising enterprises in which a reasonable good or service is being distributed, there is a rational expectation of gain to members even if they sell no franchises. Potential members may buy into the enterprise purely on the basis of this expectation. Still, it is clear that a different kind of saturation effect is important. The owner of a fast food restaurant may find that he has opened in an area which already has more such establishments than the pool of potential customers. But a careful marketing analysis will be enormously helpful in avoiding this kind of snafu. The primary saturation effect is not caused by a lack of potential purchasers of fast food restaurants but by an absence of customers. On the other hand, there is little doubt that many franchising operations infuse in potential members the idea that their main profit will be realized by selling distributorships. Indeed, many such operations are *de facto* pyramid operations. Thus, it would appear to be impossible for the government to come up with a nonstifling definition of pyramid clubs which could not be circumvented by simply providing, in addition to the recruiting license, some modest good or service (numbered "collectors' item" bronze paper weights should work nicely). The old maxim of *caveat emptor* would appear to be the best protection for the public.

The model of a pyramid club is an example of epidemic structure, although no transmission of germs is involved. Nor should the term "epidemic" be considered always to have negative connotations. It simply has to do with the ability of one population to recruit, willfully or otherwise, members of another population into its ranks at a self-sustaining rate.

2.6. A SIMPLE MODEL OF AIDS

The use of models is considered to be speculative and dangerous by some. There are many cases, however, where a failure to carry out modeling can have catastrophic results. After some decades without the onslaught on the developed nations of a fatal contagious disease with a high rate of occurrence, America became aware in the early 1980s of a mysterious "new" disease, which has been named acquired immune deficiency syndrome. Serious scientists have compared it with some of the epidemics that plagued Europe in an earlier age: cholera, the bubonic plague, and smallpox. Death totals in the United States alone have been forecast in the millions. A massive program of research into the development of a vaccine and/or a cure has been instituted. By 1987, a number of polls revealed that AIDS was the item of single greatest concern to the American electorate—exceeding nuclear war, takeover by a foreign power, or hard economic times. Into the secure yuppie age has entered a new element, which most citizens had never dreamed would occur—catastrophe by epidemic. We wish to look at a simple model of the epidemic which may give us insight into the causes of the disease, its anticipated course, and some possible strategies for intelligent epidemiological intervention.

A customary approach to the control of contagious diseases in contemporary America is via medical intervention, either by preventive vaccination or by the use of antibiotics. Historically, sociological control of epidemics has been the more customary method. This has been due, in part, to the fact that vaccines were unknown before the 19th century and antibiotics before the 20th century. It is interesting to note that for some of the classical diseases which have caused epidemics in the past, for example, cholera and the bubonic plague, the world is still without either effective vaccination or cure. Based on years of experience, it would appear that sociological control of an epidemic, rather than one based on an anticipated vaccine or cure, should be the first line of defense against a contagious disease, particularly a "new" one.

In the case of some ancient peoples, a large portion of the system of laws dealt with the means of sociological control of epidemics. For example, it should be noted that the 13th and 14th chapters and half of the 15th chapter of Leviticus (131 verses) are dedicated for the sociological control of leprosy. We might contrast this with the fact that the often mentioned dietary (kosher) laws receive only one chapter, the 11th, with a total of 47 verses.

The notion that epidemics can always be controlled by a shot or a pill rather than by the generally more painful sociological methods caused much human suffering even before AIDS. For example, First World medicine has largely displaced isolation as a control for leprosy in the Third World. Because the methods have been less effective in practice than hoped, we have the spectacle in some countries of three generations of a family sharing the same roof and the disease of leprosy.

Only in the 1980s have we (apparently) reached the level of medical control necessary to protect individuals against the effects of leprosy. But, in some sense, we have acted for some decades as though we were in possession of an antileprosy technology which we did not in actuality have.

In the case of AIDS, we see an even more difficult (of medical control) disease than leprosy. However, it does not follow that an incurable disease should produce more virulent epidemics than a disease like leprosy, which is generally regarded as crippling rather than fatal. AIDS, unlike leprosy, is not transmitted by casual contact. Consequently, we should expect that the transmission chain is much more fragile for AIDS and that relatively painless epidemiological policy can be used to stop an AIDS epidemic.

On December 1, 1987, during an NBC presidential debate, Vice President George Bush announced that the federal expenditures on AIDS research (i.e., seeking a vaccine or cure) were at levels higher than those either for cancer or heart disease. Unfortunately, a vaccine or a cure is extremely unlikely in the near future. Accordingly, we are confronted with a disease with a 100% fatality record and a per patient medical cost (using the present heroic intervention) in the $100,000/case range. We must ask the question of whether the main thrust of attack on the AIDS epidemic can be deemed optimal or even intelligent.

Let us consider an argument, developed by Thompson (1984), when the disease was first beginning to produce deaths in the several hundreds. First, we can determine the probability that a random infective will transmit the disease to a susceptible during a time interval $[t, t + \Delta t]$.

$$P(\text{transmission in } [t, t + \Delta t]) = k\alpha\Delta t \frac{X}{X + Y}; \tag{2.6.1}$$

where k = number of contacts per month,
α = probability of contact causing AIDS,
X = number of susceptibles,
Y = number of infectives.

To get the expected total increase in the infective population during $[t, t + \Delta t]$, we multiply the above by Y, the number of infectives:

$$\Delta E(Y) = YP(\text{transmission in } [t, t + \Delta t]). \tag{2.6.2}$$

For large populations, we can assume, under fairly general conditions, that the expected total change in Y is very nearly equal to a deterministic Y; that is,

$$\Delta E(Y) \approx \Delta Y. \tag{2.6.3}$$

Letting Δt go to zero, this yields immediately

$$\frac{dY}{dt} = \frac{k\alpha XY}{X + Y} \quad \text{and} \quad \frac{dX}{dt} = -\frac{k\alpha XY}{X + Y}. \tag{2.6.4}$$

We must also allow for immigration into the susceptible population, λ, and emigration, μ, from both the susceptible and infective populations and for marginal increase in the emigration from the infective population due to AIDS, γ, from sickness and death. Thus, we have the improved differential equation model

$$\frac{dY}{dt} = \frac{k\alpha XY}{X+Y} - (\gamma + \mu)Y \quad \text{and} \quad \frac{dX}{dt} = -\frac{k\alpha XY}{X+Y} + \lambda - \mu X, \tag{2.6.5}$$

where γ = AIDS death rate,
λ = immigration rate,
μ = emigration rate,

and all time rates are in months. For early stages of the disease, $X/(X + Y) \approx 1$. Accordingly, we may write the approximation

$$\frac{dY}{dt} \approx (k\alpha - \mu - \gamma)Y. \tag{2.6.6}$$

This gives us the solution

$$Y = Y(0)\exp[(k\alpha - \mu - \gamma)t]. \tag{2.6.7}$$

Now, we shall use some rough guesses for some of the parameters in the equations above. We shall assume that, absent AIDS, the total target population is 3,000,000. We shall assume that an individual stays in this population an average of 15 years (yielding $\mu = 1/(15 \times 12) = 0.00556$). We shall use as the average time an infective remains sexually active 10 months (yielding $\gamma = 0.1$). To maintain the population of 3,000,000 (absent AIDS) then, we require

$$\frac{dX}{dt} = \lambda - \mu X = 0, \tag{2.6.8}$$

or $\lambda = 16{,}666$. Now, if we combine these figures with early death data from AIDS, we can use the approximation for Y to obtain an estimate for $k\alpha \approx 0.263$. Considering that some values of k which have been variously reported range up to well over 100 per month, we are struck by the incredibly low probability of infection per contact (α) between an infective and a susceptible. This rate is exceptionally low, probably well below .01. This is very low, when compared to other venereal diseases, where values of this probability in excess of .2 are not unusual. It is this very low probability of transmission of AIDS which distinguishes its epidemiological properties and gives the key to the current AIDS epidemic. The low value of α indicates virus in some quantity is required for transmission of the disease in most cases. This means that

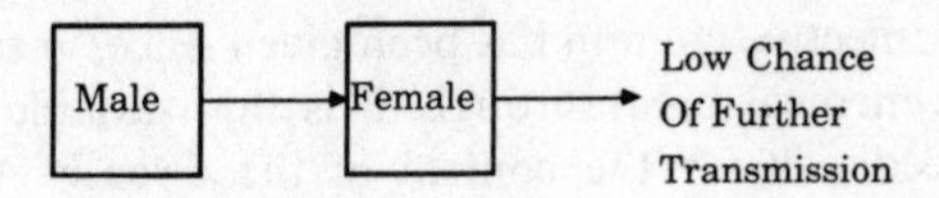

Figure 2.6.1. Heterosexual transmission pattern.

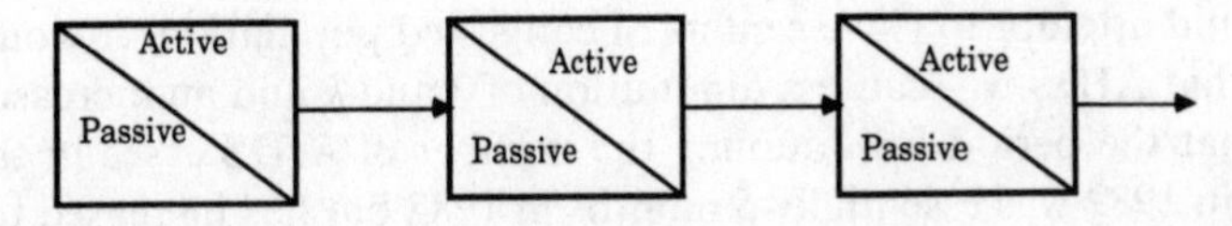

Figure 2.6.2. Homosexual transmission pattern.

transmission is much more likely from the active to the passive partner. In Figure 2.6.1, we note that this essentially "deadends" the disease in heterosexual transmission.

In the case of homosexual transmission, where participants typically play both active and passive roles, the situation is quite different (Figure 2.6.2).

Table 2.6.1 shows predicted and observed AIDS figures using the estimated parameters. Now, using the somewhat smaller $k\alpha$ value of .25 and an initial infective population of 2000, we come up with the projections given in Table 2.6.2, making the assumption that things continue with the parameter values above.

Table 2.6.1. AIDS Cases

Date	Actual	Predicted
May 82	255	189
August 82	475	339
November 82	750	580
February 83	1150	967
May 83	1675	1587

Table 2.6.2. Projections of AIDS with $k\alpha = 0.25$

Year	Cumulative Deaths	Fraction Infective
1	6,434	.004
2	42,210	.021
3	226,261	.107
4	903,429	.395
5	2,003,633	.738
10	3,741,841	.578
15	4,650,124	.578
20	5,562,438	.578

The fraction infective column has been given since, in the absence of state intervention or medical breakthrough, it is this variable that provides the (sociological) feedback for the control of the disease. Any visibility of a loathsome and fatal disease in the proportion range of 1% of the target population will almost certainly cause members of that population to consider modifying their membership in it. In the days of plague in western Europe, one could attempt to leave centers of congested population. It would appear likely that AIDS will cause a diminution of λ and k and an increase of μ. We note that the period for doubling the number of AIDS cases in the United States in 1982 was essentially 5 months in 1982 but has increased to nearly a year at present.

Let us consider, for example, the effect of diminishing k. We note that in the early stages of the disease, an equilibrium value of $k\alpha = .1056$ is obtained. At this value, with all other parameters held constant, the total body count after 20 years is 47,848 with a fraction of infectives quickly reaching 0.000668. Now, let us suppose that fear reduces k to 20% of its present value, by the use of condoms and some restraint in activity. Then, Table 2.6.3 shows that the disease quickly retreats into epidemiological insignificance.

But, let us suppose that a promiscuous fraction, p, retains a $k\alpha$ value τ times that of the less promiscuous population. Our model becomes

$$\frac{dY_1}{dt} = \frac{k\alpha X_1(Y_1 + \tau Y_2)}{X_1 + Y_1 + \tau(Y_2 + X_2)} - (\gamma + \mu)Y_1, \tag{2.6.9}$$

$$\frac{dY_2}{dt} = \frac{k\alpha\tau X_2(Y_1 + \tau Y_2)}{X_1 + Y_1 + \tau(Y_2 + X_2)} - (\lambda + \mu)Y_2,$$

$$\frac{dX_1}{dt} = -\frac{k\alpha X_1(Y_1 + \tau Y_2)}{X_1 + Y_1 + \tau(Y_2 + X_2)} + (1 - p)\lambda - \mu X_1,$$

$$\frac{dX_2}{dt} = -\frac{k\alpha\tau X_2(Y_1 + \tau Y_2)}{X_1 + Y_1 + \tau(Y_2 + X_2)} + p\lambda - \mu X_2.$$

Table 2.6.3. Projections of AIDS with $k\alpha = 0.05$

Year	Cumulative Deaths	Fraction Infective
1	1751	.00034
2	2650	.00018
3	3112	.00009
4	3349	.00005
5	3471	.00002
10	3594	.000001

Let us examine conditions under which the epidemic is not sustainable if, starting with no infectives, a small number of infectives is added. We note that if $Y_1 = Y_2 = 0$, the equilibrium values for X_1 and X_2 are $(1-p)(\lambda/\mu)$ and $p(\lambda/\mu)$, respectively. Then, expanding the right-hand sides of the two equations in Maclaurin series and neglecting terms of $O(Y_1^2)$ and $O(Y_2^2)$, we have (using lowercase symbols for the perturbations from 0)

$$\frac{dy_1}{dt} = \left(\frac{k\alpha(1-p)}{1-p+\tau p} - (\gamma+\mu)\right) y_1 + \frac{k\alpha(1-p)}{1-p+\tau p} y_2, \tag{2.6.10}$$

$$\frac{dy_2}{dt} = \frac{k\alpha\tau p}{1-p+\tau p} y_1 + \left(\frac{k\alpha\tau^2 p}{1-p+\tau p} - (\gamma+\mu)\right) y_2.$$

The solutions to a system of the form

$$\frac{dy_1}{dt} = ay_1 + by_2, \tag{2.6.11}$$

$$\frac{dy_2}{dt} = cy_1 + dy_2$$

are given by

$$y_1(t) = c_1 e^{r_1 t} + c_2 e^{r_2 t}, \tag{2.6.12}$$

$$y_2(t) = c_1 \frac{r_1 - d}{c} e^{r_1 t} + c_2 \frac{r_2 - d}{c} e^{r_2 t},$$

where

$$r_1 = \frac{a + d + \sqrt{(a+d)^2 - 4(ad - bc)}}{2}, \tag{2.6.13}$$

$$r_2 = \frac{a + d - \sqrt{(a+d)^2 - 4(ad - bc)}}{2}.$$

In order that y_1 and y_2 go to zero, and hence the disease be not sustained, we require r_1 and r_2 to be negative. Substituting in (2.6.13) from (2.6.10), we note that this is achieved if

$$\frac{k\alpha}{\gamma + \mu} < \frac{\tau p + 1 - p}{\tau^2 p + 1 - p}. \tag{2.6.14}$$

For a population with $\tau = 1$ or $p = 0$, we note that the average contact–transmission rate is simply $k\alpha$ and the disease is not sustained if

$$k\alpha < \gamma + \mu. \tag{2.6.15}$$

For $\tau > 1$ and $p > 0$, however, the average contact–transmission rate is $k\alpha[(1 - p) + \tau p]$.

One measure of enhancement of marginal increase in disease sustainability due to a subpopulation exhibiting greater sexual activity than the average is given by the ratio Q of the overall contact–transmission average required to sustain the disease in the homogeneous case divided by that for a population with a relatively high activity subpopulation:

$$Q = \frac{1 - p + \tau^2 p}{(1 - p + \tau p)^2}. \tag{2.6.16}$$

Note that Q does not depend on k, α, γ, μ, or λ. Moreover, it can be shown (Thompson, 1988) that when time delays between disease transmission and infectivity are included in the model, equation (2.6.16) still holds. In Figure 2.6.3, we exhibit a graph of Q for various activity multipliers τ and proportions p of the more active subpopulation.

We note how dramatically the presence of a small subpopulation with significantly higher sexual activity than that of the majority of the group enhances sustainability of the epidemic. The introduction of the bathhouses caused the same effect as a doubling (or more) of the contact rate and apparently caused a crossing of the epidemic threshold. Consequently, we note

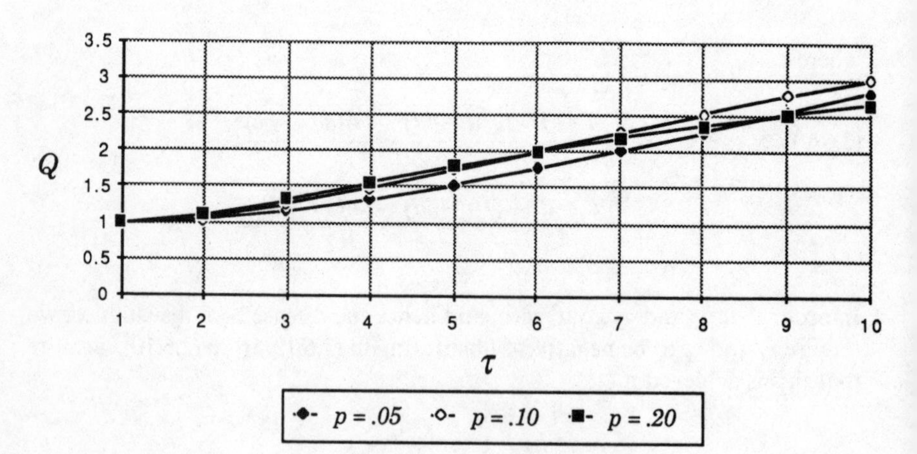

Figure 2.6.3. Multipliers of sustainability.

Table 2.6.4. Projections of AIDS with $k\alpha = .05$, $\tau = 5$, and $p = .10$

Year	Cumulative Deaths	Fraction Infective
1	2,100	.0005
2	4,102	.0006
3	6,367	.0007
4	9,054	.0008
5	12,274	.0010
10	40,669	.0020
15	105,076	.0059
20	228,065	.0091

the very real possibility that, had the public health authorities in the various municipalities in the United States simply revoked, early on, the licenses of bathhouses and other establishments which facilitate very high contact rate activity of anal intercourse, the disease of AIDS might never have reached epidemic proportions in this country. Even more intriguing is the question as to whether absent the liberalized licensing regulations of such establishments promulgated in the late 1970s and early 1980s, AIDS would ever have developed in significant numbers. It is interesting to note that many leaders of the gay community requested bathhouse closings from the early days of the epidemic but were generally rebuffed by the municipal public health authorities (Shilts, 1987, pp. 328, 431, 441–443, 454, 464).

In Table 2.6.4, we consider the case where $k\alpha = .05$, $\tau = 5$, and $p = .1$. We note how the presence of even a small promiscuous population can stop the demise of the epidemic. But, if this proportion becomes sufficiently small, then the disease is removed from an epidemic to an endemic situation, as we see in Table 2.6.5 with $p = .05$ and all other parameters the same as above. The

Table 2.6.5. Projections with $k\alpha = .05$, $\tau = 5$, and $p = .05$

Year	Cumulative Deaths	Fraction Infective
1	1917	.00043
2	3272	.00033
3	4344	.00027
4	5228	.00022
5	5971	.00019
10	8263	.00008
15	9247	.00003
20	9672	.00002

Table 2.6.6. Projections of AIDS with $k\alpha = .02$, $\tau = 16$, and $p = .1$

Year	Cumulative Deaths	Fraction Infective
1	2,184	.0007
2	6,536	.0020
3	20,583	.0067
4	64,157	.0197
5	170,030	.0421
10	855,839	.0229
15	1,056,571	.0122
20	1,269,362	.0182

dramatic effect of a small promiscuous population may be considered in the case where 90% of the population has a $k\alpha$ of .02 and 10% has a $k\alpha$ of .32. This gives a population with an overall $k\alpha$ of .05. If this low value is maintained uniformly across the population, then we have seen that the disease quickly dies out. But consider the situation when the mix is given as in Table 2.6.6. Thus, it is quite clear that, with the same overall group average contact rate, a population exhibiting homogeneous activity is much more resistant to sustaining an epidemic than one with a small high contact rate subpopulation.

It is now clear that the presence of AIDS in the human population is of long standing. We have AIDS-producing virus in U.S. human blood specimens going back at least to 1969, in specimens from human blood in Zaire at least as far back as 1959. Furthermore, we now have Guyader's (1987) virological evidence that AIDS is of ancient origin. At least in the case of AIDS, the virologist's maxim that all diseases are old diseases seems validated.

Since AIDS is not a new disease, then we ought to ask what has changed in order that an endemic disease has now reached epidemic proportions. After all, we should recall that Belgian colonial troops went back and forth from Zaire to Belgium for nearly 100 years. If there was ever a significant outbreak of AIDS in Belgium during the colonial epoch, no one has discovered the fact. It seems most likely that the reason is that the high contact rates that characterize some segments of the homosexual communities that exist in some American cities have never occurred before in the history of the world.

Naturally, there is concern as to whether AIDS can be sustained heterosexually in the United States. The main argument in support of such a possibility is the fact that the concurrent Central African AIDS epidemic is apparently almost completely a heterosexual phenomenon. How much of the African transmission is due to nonsexual modes, for example, local health officials dispensing medical care with unsterilized needles and surgical instruments, and how much is due to untreated lesions from other venereal diseases dramatically enhancing the possibility of female to male transmission (an essential mode if the disease is to be transmitted heterosexually) are a matter of conjecture. But we know that economic hard times have seriously stressed the quality of medicine in many African countries, and the two modes of transmission mentioned above may explain the heterosexual AIDS epidemic there. If so, it is very likely that the AIDS epidemic in Africa is the indirect consequence of conditions of economic decline, which are not likely to be matched anywhere in the United States. Of the 45,425 adult American cases of AIDS listed by the CDC on November 16, 1987, slightly less than 4% were effected by heterosexual transmission. Of the 42,208 male cases, less than 2% were heterosexually transmitted. Interestingly, of the 3217 female cases, 29% were the result of heterosexual transmission (the rest being by IV drug use and blood transfusion). The argument that female-to-male transmission is unlikely except in special circumstances like those mentioned above still squares with the data over 5 years into the epidemic. The indications are that conditions do not exist in the United States which can produce a stand-alone heterosexual epidemic. However, any failure of the American public health community to provide effective health care delivery to stop epidemics of other venereal diseases than AIDS in communities of the disadvantaged could facilitate a heterosexual AIDS epidemic in such communities.

One prediction about AIDS is that there is a "Typhoid Mary" phenomenon. That means that the actual transmission rate is much higher than had been supposed, but only a fraction of the infected develop the disease quickly. Another fraction become carriers of the disease without themselves actually developing the physical manifestations of the disease, except possibly after a long interval of time. To see the effects of such a phenomenon, let us suppose $k\alpha = .05$, but 50% of those who contract the disease have a sexually active life expectancy of 100 months instead of only 10. See Table 2.6.7.

Such a disastrous scenario is naturally made much worse as we increase the fraction of those with the long sexually active life expectancy. For example, if this proportion is 90%, we have the results given in Table 2.6.8.

Table 2.6.7. Projections of AIDS with $k\alpha = .05$ and Half of the Infectives with $\gamma = 0.01$

Year	Cumulative Deaths	Fraction Infective
1	1,064	.00066
2	1,419	.00075
3	2,801	.00089
4	3,815	.00110
5	5,023	.00130
10	16,032	.00330
15	44,340	.00860
20	115,979	.02210

Table 2.6.8. Projections of AIDS with $k\alpha = .05$ and 90% of the Infectives Having $\gamma = 0.01$

Year	Cumulative Deaths	Fraction Infective
1	457	.0094
2	1,020	.0013
3	1,808	.0020
4	2,943	.0028
5	4,587	.0041
10	32,911	.0260
15	194,154	.1441
20	776,146	.4754

If the Typhoid Mary phenomenon is an actuality, then the effect of AIDS is likely to be catastrophic indeed. (Note that no presence of a promiscuous subpopulation is necessary to cause this catastrophic scenario.) However, this would imply that AIDS was a new disease, since such a phenomenon would have sustained the epidemic in an earlier time. As pointed out above, we have evidence that AIDS is an old disease. Furthermore, a significant Typhoid Mary phenomenon would appear to work against the aforementioned continual lengthening of the doubling time for observed cases of the disease.

Let us suppose we lived in some strange society where the public health establishment decided to maximize the fatalities due to a "new" contagious disease. What strategy would be "optimal" to achieve this result? It would be not to stress the epidemiological transmission chain, but rather to rely on the development of a cure or vaccine of the disease. A great deal of publicity would be given as to the large amounts of money being spent in attempting to find such cures and vaccines. Public officials would be encouraged to make frequent speeches as to their commitment to finding a cure, whatever the cost. In short, insofar as the public health establishment dealt with the epidemic (rather than the disease), it would be to eliminate fear of the disease.

Unfortunately, with the best of intentions and a high sensitivity to civil liberties, American public health policy in dealing with AIDS has come perilously close to the strategy mentioned above. With all the hundreds of millions already spent in the search for a cure, we have none and are now warned that we are probably years from finding one. The even greater portion of funds expended on finding a preventive vaccine is of even less public health utility in ending the current epidemic. If such a vaccine were found very soon (an event predicted by no responsible medical researcher), the vast numbers of individuals already infected by the disease would still perish.

If there is a bright side, it seems to be that fear-driven behavior modification has probably already broken the back of the AIDS epidemic in the United States. The overwhelming majority of those who will succumb to the disease in this century probably already have the virus in their systems, and, in the epidemiological sense, have acquired the disease. The most important elements in AIDS which will cause its essential elimination are the low value of α and fear of the disease. Predictably, as the diseased fraction of the target population increases, k will decrease, λ will decrease, p will decrease, and μ will increase. Indeed, these trends have been under way for some time. We note again that the doubling time of AIDS, which was 5 months in 1982, has already increased to 1 year.

In the long run, members of a target population will tend to take the necessary steps to lessen their participation in an epidemic. So it has been with AIDS. But it is the function of a public health establishment to shortcircuit the painful path dependent on experience of members of the target population. Had public health authorities vigorously attacked the transmission chain early on by shutting down high contact inducing establishments, the present AIDS holocaust in the gay community might well have been avoided. By simply satisfying itself with collecting statistics rather than implementing sociological controls for eliminating an epidemic, the Centers for Disease Control have behaved very much like a fire department that contents itself with merely recording the number of houses burned down.

As this book goes to press in 1988, not one of the several federal or state commissions dealing with the AIDS epidemic has yet to recommend that the gay bathhouses be closed.

References

CDC (1983). Update: Acquired immunodeficiency syndrome (AIDS), *United States Morbidity and Mortality Weekly Report,* **32,** 389–391.

CDC (1986). Update: Acquired immunodeficiency syndrome (AIDS), *United States Morbidity and Mortality Weekly Report,* **35,** 17–20.

CDC (1987). Update: Acquired immunodeficiency syndrome (AIDS), *United States Morbidity and Mortality Weekly Report,* **36,** 807.

Crewdson, John (1987). Doctors say AIDS virus caused teen's '69 death, *Houston Chronicle,* October 25, 1987, Section 1, 4 (reprinted from the *Chicago Tribune*).

Guyader, Mireille; Emerman, Michael; Sonigo, Pierre; Clavel, François; Montagnier, Luc and Alizon, Marc. (1987). Genome organization and transactivation of the human immunodeficiency virus type 2, *Nature,* **326,** 662–669.

Shilts, Randy (1987). *And The Band Played On: Politics, People, And The AIDS Epidemic,* St. Martin's, New York.

Thompson, James R. (1984). Deterministic versus stochastic modeling in neoplasia, *Proceedings of the 1984 Summer Computer Simulation Conference,* pp. 822–825.

Thompson, James R. (1988). AIDS: The mismanagement of an epidemic, to appear in *Mathematical Population Dynamics.*

CHAPTER THREE

Simulation and the Coming Qualitative Change in Scientific Modeling

3.1. SIMULATION-BASED TECHNIQUES FOR DEALING WITH PROBLEMS USUALLY APPROACHED VIA DIFFERENTIAL EQUATION MODELING

It is interesting to recall that von Neumann was motivated to conceive of and build the first serious digital computer as a device for handling simulation algorithms which he had formulated for dealing with problems in nuclear engineering. Ideally, if we are dealing with problems of heat transfer, neutron flux, and so on, in regular and symmetrical regions, the classical 19th and early 20th century differential–integral-difference equation formulations can be used. However, if the regions are complicated, if indeed we are concerned about a maze of pipes, cooling vessels, rods, and so on, the closed form solutions are not available. This means many person years to come up with all the approximation theoretic quadrature calculations to ensure that a satisfactory plant will result if the plans are implemented. von Neumann noted that if large numbers of simple repetitive computations could readily be performed by machine, then a method could be constructed which would serve as an alternative to quadrature. For example, if we wish to compute

$$\int_0^1 g(x)\,dx, \tag{3.1.1}$$

we might be so fortunate as to have a $g(\cdot)$ whose antiderivative we know in closed form. But if we had one whose antiderivative we did not have (e.g.,

$g(x) = \exp(-x^2)$), we might employ the simplest quadrature formula, the trapezoidal rule; that is,

(3.1.2)
$$\int_0^1 g(x)\,dx \approx \left[g(0) + 2g\left(\frac{1}{n}\right) + 2g\left(\frac{2}{n}\right) + \cdots + 2g\left(\frac{n-1}{n}\right) + g(1) \right] \frac{1}{2n}.$$

Unless g is unusually ill-behaved, we should be able to make do with some such quadrature rule for a one-dimensional problem even using the formula by hand. Now let us suppose that the integration is not one-dimensional but is over the five-dimensional unit hypercube. Then, the quadrature formulas become much less attractive even if computing is relatively cheap. For example, suppose that we have the capabilities for evaluating $g(\cdot)$ at 100,000 points. This allows us 10 quadrature levels in each of the five dimensions. Since these levels will generally be chosen with a fair amount of regularity (e.g., equally spaced) and since the function g is also selected by someone with at least a subconscious intent of picking something regular, we might well discover that the quadrature approach completely missed regions of some interest, for example, points where the function becomes very large.

In such a situation, it would probably be much safer to select points drawn randomly from the hypercube. Among the virtues of such a Monte Carlo approach is the fact that such randomized quadrature points would be less likely to be in synchrony with g in some bizarre fashion. For example, in one dimension, suppose we wished to evaluate

$$\int_0^1 |\sin(1000\pi x)|\,dx. \tag{3.1.3}$$

It is easy to see how a 50 point or a 100 point or a 200 point or a 500 point or a 1000 point quadrature formula could give us an apparent value of (3.1.3) to be zero. If the points were chosen irregularly, then this sort of distortion would happen much less frequently. Clearly, as the dimensionality of the integration increases, as we must coarsen any regular grid substantially, such problems as that shown above become more and more troublesome for regular grids: hence, the notion of the randomized quadrature approach sometimes designated as "Monte Carlo."

Much has been written about clever means for generating apparently random numbers on the computer. Comments have been made to the effect that we should count ourselves lucky that the rather crude random number generation procedure used by von Neumann did not lead to catastrophe. More recently, the generator RANDU, which was developed by IBM and was at one time the most used of all random generators, tended to produce random

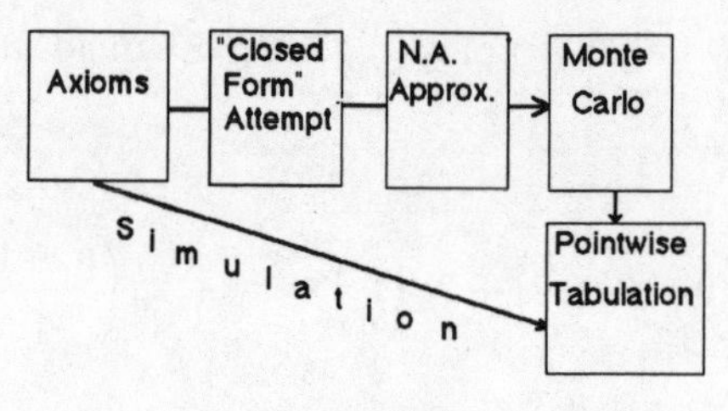

Figure 3.1.1

numbers that fell in rather far apart lattices even in dimension two. It is perhaps interesting to note that there appear to be no instances where researchers using generators which cycled too soon or gave rise to noticeable lattice structure came up with results that produced catastrophes.

In the long run, the quadrature issue, which Monte Carlo was largely developed to address, is rather unimportant when compared to the much more important issue of direct simulation. To make a distinction between Monte Carlo and simulation, let us consider the following two paradigms shown in Figure 3.1.1. In the upper flowchart, we note a traditional means of coping with the numerical results of a model. We start out with axioms at the microlevel which are generally easily understood. For example, one such axiom might be that a particle starts at a particular point and moves step by step in three space according to specified laws until it collides with and is absorbed by a wall. Another might be that each tumor cell at the moment of its formation (as one of two daughter cells following division of the mother cell) breaks off from the tumor mass and forms a separate metastatic clone with constant probability (for each cell).

Naturally, the task of dealing with each specific gas molecule out of a total of, say, 10^9 molecules or keeping a record of each of 10^{10} tumor cells is a hopeless task. Thus, investigators in the 19th century quite naturally and correctly were led to means for summary information about the gas molecules or about the metastatic progression of the tumor.

So then, the upper path gives the traditional paradigm for solving such problems. We start with axioms that are frequently agreed to by most of the investigators in the field. These are transformed into a differential–integral-difference equation type of summary model. Next, a generally pro forma attempt is made to arrive at a closed form solution—that is, a representation that can be holistically comprehended by an intelligent observer and that lends itself to precise numerical evaluation of the dependent variables as we change the parameters of the model and the independent variables. This attempt is generally unsuccessful and leads only to some nonholistic quadrature-like setup for numerical evaluation of the independent variables. If the dimensionality of the quadrature is greater than two, then the user moves rather quickly to a "random quadrature" Monto Carlo approach.

The object of this chapter is to indicate how we now have the computer speed to use the algorithm in the lower part of the diagram. Namely, we can now frequently dispense with the traditional approach and use one that goes directly from the microaxioms to pointwise evaluation of the dependent variables. The technique for making this "great leap forward" is, in principle, simplicity itself. Simulation simply carries out that which would earlier have been thought to be absurd, namely, follow the progress of the particles, the cells, whatever. We do not do this for all the particles or cells, but only for a representative sample. We still do not have the computer speed to deal with 10^{10} particles or cells; but we can readily deal with, say, 10^4 or 10^5. For many purposes, such a size is more than sufficient to yield acceptable accuracy. Among the advantages of a simulation approach is principally that it enables us to eliminate time-consuming and artificial approximation theoretic activities and to spend our time in more useful pursuits. More importantly, simulation will enable us to deal with problems that are so complex in their "closed form" manifestation that they are presently attacked only in ad hoc fashion. For example, econometric approaches are generally linear—not because such approaches are supported by microeconomic theory but because the complexities of dealing with the aggregate consequences of the microeconomic theory are so overwhelming. Similarly, in mathematical oncology, the use of linear models is motivated by the failure of the natural branching process models to lead to numerically approximateable "closed forms." It is the belief of the author that the current generation of 32 bit chips can bring about the real computer revolution and change fundamentally the ways in which we approach the task of modeling and problem solving. Indeed, as is frequently the case, hardware seems to lead software by a decade. We have had the hardware capabilities for a long time to implement all the techniques covered in this chapter. But the proliferation of fast computing to the desktop will encourage private developers to develop simulation-based procedures for a large and growing market of users who need to get from specific problems to useful solutions in the shortest time possible. We now have the ability to use the computer not as a fast calculator but as a device that changes fundamentally the process of going from the microaxioms to the macrorealization.

To explicate this "simulation revolution" concept, we shall begin with an apparently Monte Carlo based approach given by Lattes (1969). Let us consider the old problem of gambler's ruin.

Suppose there are two gamblers, A and B, who start to gamble with stakes x and $b - x$, respectively. At each round, each gambler puts up a stake of h dollars. The probability that A wins a round is p; the probability that B wins a round is $q = 1 - p$. We wish to compute the probability that A ultimately wins the game. Let us define $v(x, t)$ as the probability that A wins the game starting with capital x on or before the tth round. Similarly, $u(x, t)$ is the probability that B wins the game with his stake of $b - x$ on or before the tth round. Let $w(x, t)$ be the probability the game has not terminated by the tth round.

First, we note that each of the three variables v, u, and w is bounded below by zero and above by one. Furthermore, u and v are nondecreasing in t, while w is nonincreasing in t. Thus, we can take limits of each of these as t goes to infinity. We shall call these limits $v(x)$, $u(x)$, and $w(x)$, respectively.

A fundamental recursion, which will be the basis for practically everything we do in this section, is given by

$$v(x, t) = pv(x + h, t - \lambda) + qv(x - h, t - \lambda). \tag{3.1.4}$$

This simply means that the probability A wins the game on or before the tth round is given by the probability he wins the first round and then ultimately wins the game with his new stake of $x + h$ in $t - \lambda$ rounds plus the probability he loses the first round and then wins the game in $t - \lambda$ rounds with his new stake of $x - h$. Here we have used a time increment of λ.

Taking limits in (3.1.4), we have

$$(p + q)v(x) = pv(x + h) + qv(x - h). \tag{3.1.5}$$

Rewriting (3.1.5), we have

$$p[v(x + h) - v(x)] = q[v(x) - v(x - h)]. \tag{3.1.6}$$

Let us make the further simplifying assumption that $b = Nh$. Then

$$v(\{n + 1\}h) - v(nh) = \frac{q}{p}[v(nh) - v(\{n - 1\}h)]. \tag{3.1.7}$$

We note that $v(0) = 0$ and $v(Nh) = 1$. Then writing (3.1.7) in extenso, we have

$$v(nh) - v(\{n - 1\}h) = \frac{q}{p}[v(\{n - 1\}h) - v(\{n - 2\}h)]$$

$$v(\{n - 1\}h) - v(\{n - 2\}h) = \frac{q}{p}[v(\{n - 2\}h) - v(\{n - 3\}h)]$$

$$\vdots$$

$$v(2h) - v(h) = \frac{q}{p}[v(h) - v(0)] = \frac{q}{p}v(h).$$

Substituting up the ladder, we have

$$v(nh) - v(\{n-1\}h) = \left(\frac{q}{p}\right)^{n-1} v(h). \tag{3.1.8}$$

Substituting (3.1.8) in the extenso version of (3.1.7), we have

$$\begin{aligned} v(nh) - v(\{n-1\}h) &= \left(\frac{q}{p}\right)^{n-1} v(h) \\ v(\{n-1\}h) - v(\{n-2\}h) &= \left(\frac{q}{p}\right)^{n-2} v(h) \\ &\vdots \\ v(h) &= v(h). \end{aligned} \tag{3.1.9}$$

Adding (3.1.9), we have

$$v(nh) = \left[1 + \left(\frac{q}{p}\right) + \left(\frac{q}{p}\right)^2 + \cdots + \left(\frac{q}{p}\right)^{n-1}\right] v(h). \tag{3.1.10}$$

Recalling that $v(Nh) = 1$, we have

$$1 = \left[1 + \left(\frac{q}{p}\right) + \left(\frac{q}{p}\right)^2 + \cdots + \left(\frac{q}{p}\right)^{N-1}\right] v(h). \tag{3.1.11}$$

Thus, we have

$$v(nh) = \frac{1 + (q/p) + (q/p)^2 + \cdots + (q/p)^{n-1}}{1 + (q/p) + (q/p)^2 + \cdots + (q/p)^{N-1}}. \tag{3.1.12}$$

For $p = q = .5$, this gives,

$$v(nh) = \frac{n}{N}, \quad \text{or} \quad v(x) = \frac{x}{b}. \tag{3.1.13}$$

Otherwise, multiplying (3.1.12) by $(1 - p/q)/(1 - p/q)$, we have

$$v(nh) = \frac{1 - (q/p)^n}{1 - (q/p)^N}, \quad \text{or} \quad v(x) = \frac{1 - (q/p)^{x/h}}{1 - (q/p)^{b/h}}. \tag{3.1.14}$$

Now, by symmetry,

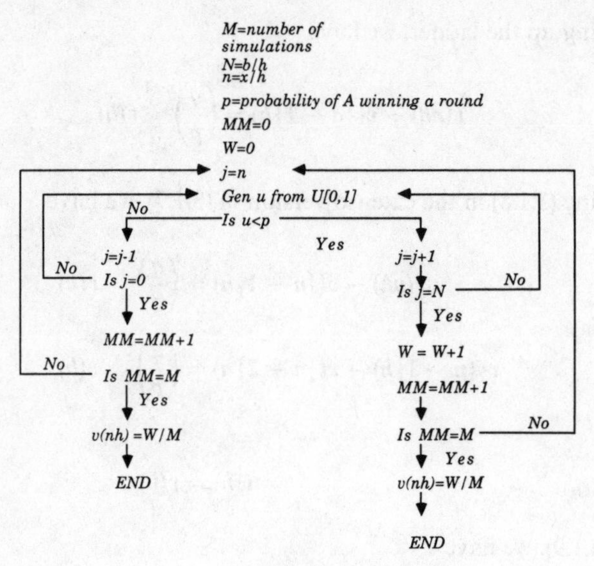

Figure 3.1.2

$$u(x) = \frac{(q/p)^{x/h} - (q/p)^{b/h}}{1 - (q/p)^{b/h}}. \tag{3.1.15}$$

From (3.1.14) and (3.1.15), we have

$$v(x) + u(x) = 1. \tag{3.1.16}$$

Consequently, $w(x) = 0$; that is, the game must terminate with probability 1. Thus, we can use a simulation to come up with reasonable estimates of the probability of A ultimately winning the game. A flowchart of such a simulation is given in Figure 3.1.2. We note that this simulation gives us a ready means of estimating a rough 95% confidence interval for $v(x)$, namely,

$$v(x) = \frac{W}{M} \pm 2\frac{\sqrt{W(1 - W/M)}}{M}. \tag{3.1.17}$$

Naturally, since we have shown a closed form solution for the gambler's ruin problem, it would be ridiculous for us to use a simulation to solve it. It is by means of an analogy of real-world problems to the general equation (3.1.4) that simulation becomes useful.

Rewriting (3.1.6), we have

$$p\,\Delta v(x) = q\,\Delta u(x - h), \tag{3.1.18}$$

where

$$\Delta v(x) = \frac{v(x+h) - v(x)}{h}.$$

Subtracting $q\,\Delta v(x)$ from both sides of (3.1.18), we have

$$(p - q)\Delta v(x) = q[\Delta v(x-h) - \Delta v(x)], \tag{3.1.19}$$

or

$$\Delta^2 v(x) + \frac{p-q}{qh}\Delta v(x) = 0, \tag{3.1.20}$$

where

$$\Delta^2 v(x) = \frac{\Delta v(x) - \Delta v(x-h)}{h}.$$

But for h sufficiently small, this is an approximation to

$$\frac{d^2 v}{dx^2} + 2\beta\frac{dv}{dx} = 0, \tag{3.1.21}$$

where

$$\frac{p-q}{qh} = 2\beta.$$

Now, suppose we are given the boundary conditions of (3.1.21) $v(0) = 0$ and $v(b) = 1$. Then our flowchart in Figure 3.1.2 gives us a ready means of approximating the solution to (3.1.21). We simply set $p = (2\beta h + 1)/(2\beta h + 2)$, taking care to see that h is sufficiently small, if β be negative, to have p positive. To make sure that we have chosen h sufficiently small that the simulation is a good approximation to the differential equation, typically we use simulations with successively smaller h until we see little change in $v(x) \approx W/N$.

What if the boundary conditions are rather less accommodating; for example, suppose $v(0)$ and $v(b)$ take arbitrary values? A moment's reflection shows that

$$v(x) \approx \frac{W}{N}v(b) + \left(1 - \frac{W}{N}\right)v(0). \tag{3.1.22}$$

Of course, a closed form solution of (3.1.21) is readily available. We need not consider simulation for this particular problem. But suppose that we generalize (3.1.21) to the case where β depends on x:

$$\frac{d^2v}{dx^2} + 2\beta(x)\frac{dv}{dx} = 0. \tag{3.1.23}$$

Then we use our flowchart in Figure 3.1.2, except that at each step, we change p via

$$p(x) = \frac{2\beta(x)h + 1}{2\beta(x)h + 2}. \tag{3.1.24}$$

Once again, $v(x) \approx (W/N)v(b) + (1 - W/N)v(0)$. And once again, it is an easy matter to come up with an internal measure of accuracy via (3.1.17).

Naturally, it is possible to effect numerous computational efficiencies. For example, we need not start afresh for each new grid value of x. For each pass through the flowchart, we can note all grid points visited during the pass and increase the counter of wins at each of these if the pass terminates at b, the number of losses if the pass terminates at 0.

It is important to note that the simulation used to solve (3.1.21) actually corresponds, in many cases, to the microaxioms to which the differential equation (3.1.21) is a summary. This is very much the case for the Fokker–Planck equation, which we consider below.

Let us suppose we do not eliminate time in (3.1.4). We shall define the following:

$v(x, t, 0; h, \lambda) = P$(particle starting at x will be absorbed at 0 on or before $t = m\lambda$);

$v(x, t, b; h, \lambda) = P$(particle starting at x will be absorbed at b on or before $t = m\lambda$);

$V(x, t; h, \lambda) = V(0, t)v(x, t, 0; h, \lambda) + V(b, t)v(x, t, b; h, \lambda)$.

Typically, we will have time constant boundary values. We define

$$\Delta_t V(x, t; h, \lambda) = \frac{V(x, t + \lambda; h, \lambda) - V(x, t; h, \lambda)}{\lambda}, \tag{3.1.25}$$

$$\Delta_x V(x, t; h, \lambda) = \frac{V(x + h, t; h, \lambda) - V(x, t; h, \lambda)}{h},$$

$$\Delta_{xx} V(x, t; h, \lambda) = \frac{\Delta_x V(x, t; h, \lambda) - \Delta_x V(x - h, t; h, \lambda)}{h}.$$

Now, our basic relation in (3.1.4) still holds, so we have

$$V(x, t + \lambda; h, \lambda) = p(x)V(x + h, t; h, \lambda) + q(x)V(x - h, t; h, \lambda). \tag{3.1.26}$$

Subtracting $V(x, t; h, \lambda)$ from both sides of (3.1.26), we have

$$\begin{aligned}
\lambda \Delta_t V(x, t; h, \lambda) &= p(x)[V(x+h, t; h, \lambda) - V(x, t; h, \lambda)] \\
&\quad + q(x)[V(x-h, t; h, \lambda) - V(x, t; h, \lambda)] \\
&= p(x)[V(x+h, t; h, \lambda) - V(x, t; h, \lambda)] \\
&\quad - q(x)[V(x, t; h, \lambda) - V(x-h, t; h, \lambda)] \\
&= hp(x)\Delta_x V(x, t; h, \lambda) - hq(x)\Delta_x V(x-h, t; h, \lambda) \\
&= h[p(x) - q(x)]\Delta_x V(x, t; h, \lambda) \\
&\quad + hq(x)[\Delta_x V(x, t; h, \lambda) - \Delta_x V(x-h, t; h, \lambda)] \\
&= h[p(x) - q(x)]\Delta_x V(x, t; h, \lambda) + h^2 q(x)\Delta_{xx} V(x, t; h, \lambda).
\end{aligned} \tag{3.1.27}$$

Now, letting $p(x) = [\beta(x) + 2h\alpha(x)]/[2\beta(x) + 2h\alpha(x)]$ and $q(x) = 1 - p(x)$, we have

$$\begin{aligned}
\lambda \Delta_t V(x, t; h, \lambda) &= 2h^2 \frac{\alpha(x)}{2\beta(x) + 2h\alpha(x)} \Delta_x V(x, t; h, \lambda) \\
&\quad + h^2 \frac{\beta(x)}{2\beta(x) + 2h\alpha(x)} \Delta_{xx} V(x, t; h, \lambda).
\end{aligned} \tag{3.1.28}$$

Now, taking h very small with $\lambda/h^2 = \mu$, we have

$$\mu \Delta_t V(x, t; h, \lambda) = \frac{\alpha(x)}{\beta(x)} \Delta_x V(x, t; h, \lambda) + \tfrac{1}{2}\Delta_{xx} V(x, t; h, \lambda). \tag{3.1.29}$$

So the simulation, which proceeds directly from the microaxioms, yields in the limit as the infinitesimals go to zero, a practical pointwise evaluator of the usual Fokker–Planck equation, the mathematically nontractable summary of those axioms:

$$2\mu \frac{\partial V}{\partial t} = 2 \frac{\alpha(x)}{\beta(x)} \frac{\partial V}{\partial x} + \frac{\partial^2 V}{\partial x^2}. \tag{3.1.30}$$

Again, our algorithm is essentially the flowchart in Figure 3.1.2 with a time counter added on.

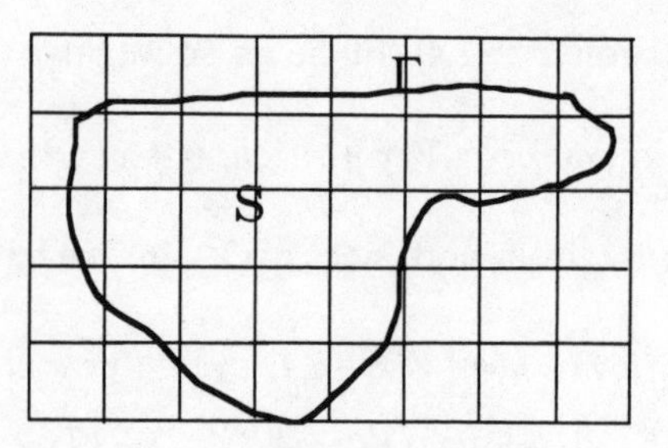

Figure 3.1.3

Let us now consider another common differential equation model of physics, that of Dirichlet. In R_k, let there be given a bounded connected region S with boundary Γ. Let there be given a function $\phi(x)$ satisfying the equation of Laplace inside S:

$$\sum_{j=1}^{k} \frac{\partial^2 \phi_j}{\partial x_j^2} = 0. \tag{3.1.31}$$

At every given boundary point Q, the values of ϕ are given explicitly by the piecewise continuous function $f(Q)$; that is,

$$\phi(x)|_\Gamma = f(Q). \tag{3.1.32}$$

For most boundaries and boundary functions, the determination of ϕ analytically is not known. The usual numerical approximation approach can require a fair amount of setup work, particularly if the dimensionality is three or greater. We exhibit in Figure 3.1.3 a simulation technique that is, in fact, an actualization of the microaxioms that frequently give rise to (3.1.31). Although our discussion is limited to R_2, the generalization to R_k is quite straightfor-

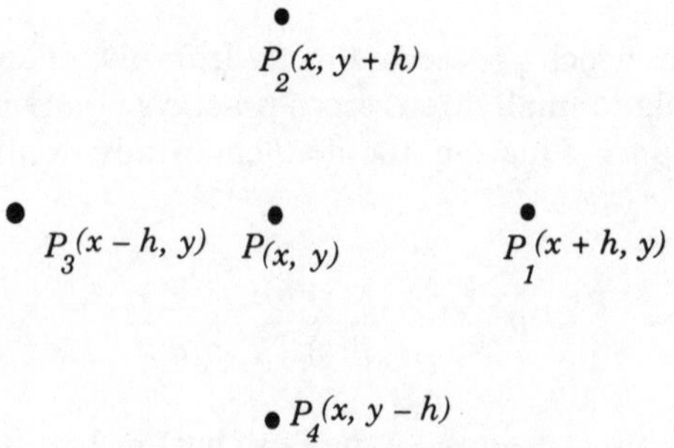

$P_2(x, y + h)$

$P_3(x - h, y)$ $P(x, y)$ $P_1(x + h, y)$

$P_4(x, y - h)$

Figure 3.1.4

ward. Let us superimpose over S a square grid of length h on a side. The points of intersection inside S nearest Γ shall be referred to as boundary nodes. All other nodes inside S shall be referred to as internal nodes.

Let us consider an internal node with coordinates (x, y) in relation to its four immediate neighbors (Figure 3.1.4). Now,

$$\left.\frac{\partial \phi}{\partial x}\right|_{(x,y)} \approx \frac{\phi(x + (h/2), y) - \phi(x - (h/2), y)}{h} \tag{3.1.33}$$

and

$$\frac{\partial^2 \phi}{\partial x^2} \approx \frac{\left.\frac{\partial \phi}{\partial x}\right|_{(x+(h/2),y)} - \left.\frac{\partial \phi}{\partial x}\right|_{(x-(h/2),y)}}{h} \approx \frac{\phi(x + h, y) + \phi(x - h, y) - 2\phi(x, y)}{h^2}. \tag{3.1.34}$$

Similarly,

$$\frac{\partial^2 \phi}{\partial y^2} \approx \frac{\phi(x, y + h) + \phi(x, y - h) - 2\phi(x, y)}{h^2}. \tag{3.1.35}$$

Laplace's equation then gives

$$0 = \frac{\partial^2 \phi}{\partial x^2} + \frac{\partial^2 \phi}{\partial y^2} \approx \frac{\phi(P_1) + \phi(P_2) + \phi(P_3) + \phi(P_4) - 4\phi(P)}{h^2}. \tag{3.1.36}$$

So

$$\phi(P) \approx \frac{\phi(P_1) + \phi(P_2) + \phi(P_3) + \phi(P_4)}{4}. \tag{3.1.37}$$

Equation (3.1.37) gives us a ready means of a simulation solution to the Dirichlet problem. Starting at the internal node (x, y), we randomly walk to one of the four adjacent points with equal probabilities. We continue the process until we reach a boundary node, say, Q_i. After N walks to the boundary from starting point (x, y), our estimte of $\phi(x, y)$ is given simply by

$$\phi(x, y) \approx \frac{\sum_{i=1}^{N} n_i f(Q_i)}{N}. \tag{3.1.38}$$

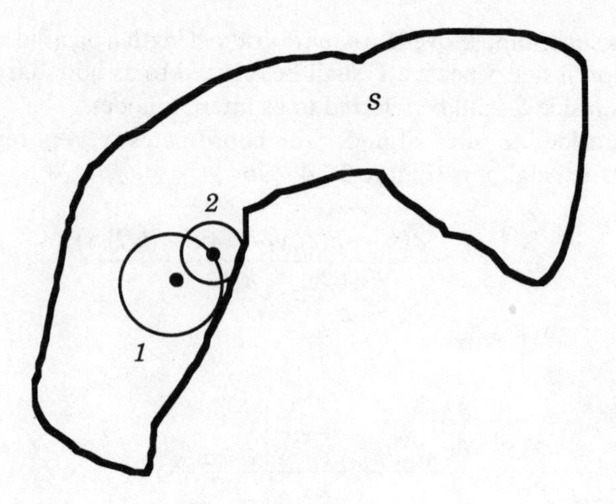

Figure 3.1.5

where n_i is the number of walks terminating at boundary node Q_i and the summation is taken over all boundary nodes.

In the above, if we wish to show ϕ contours throughout S, we can take advantage of a number of computational efficiencies. For example, as we walk from (x, y) to the boundary, we traverse $(x + h, y)$ numerous times. By incorporating the walks that traverse $(x + h, y)$ even though $(x + h, y)$ is not our starting point, we can increase the total number of walks used in the evaluation of $\phi(x + h, y)$.

Let us now consider a technique that is particularly useful if we need to evaluate ϕ at only one point in S (see Figure 3.1.5). Since it can easily be shown that $\phi(x, y)$, the solution to Laplace's equation inside S, is equal to the average of all values taken on a circle centered at (x, y) and lying inside S, we can draw around (x, y) the largest circle lying inside S and then select a point uniformly on the boundary of that circle, use it as the center of a new circle, and select a point at random on that circle. We continue the process until we arrive at a boundary node. Then, again after N walks,

$$\phi(x, y) \approx \frac{\sum_{i=1}^{N} n_i f(Q_i)}{N}. \tag{3.1.39}$$

Naturally, the above method works, using hyperspheres, for any dimension k. Again, it should be emphasized that in many cases the simulation is a direct implementation of the microaxioms which gave rise to Laplace's equation.

Let us now consider a rather general elliptic differential equation in a region S in two-dimensional space. Again, the values on the boundary Γ are given by $f(\cdot)$, which is piecewise continuous on Γ. Inside S,

$$\beta_{11}\frac{\partial^2\phi}{\partial x^2} + 2\beta_{12}\frac{\partial^2\phi}{\partial x\partial y} + \beta_{22}\frac{\partial^2\phi}{\partial y^2}\beta_{12} + 2\alpha_1\frac{\partial\phi}{\partial x} + 2\alpha_2\frac{\partial\phi}{\partial y} = 0, \tag{3.1.40}$$

where $\beta_{11} > 0, \beta_{22} > 0$, and $\beta_{22} - \beta_{12} > 0$. We consider the difference equation corresponding to (3.1.40); namely,

$$\beta_{11}\Delta_{xx}\phi + 2\beta_{12}\Delta_{xy}\phi + \beta_{22}\Delta_{yy}\phi + 2\alpha_1\Delta_x\phi + 2\alpha_2\Delta_y\phi = 0. \tag{3.1.41}$$

Although several approximations to the finite differences might be employed, one that is convenient here is given by

$$\Delta_{xy} = \frac{\phi(x+h, y+h) - \phi(x, y+h) - \phi(x+h, y) + \phi(x, y)}{h^2}, \tag{3.1.42}$$

$$\Delta_{xx} = \frac{\phi(x+h, y) + \phi(x-h, y) - 2\phi(x, y)}{h^2},$$

$$\Delta_{yy} = \frac{\phi(x, y+h) + \phi(x, y-h) - 2\phi(x, y)}{h^2},$$

$$\Delta_x = \frac{\phi(x+h, y) - \phi(x, y)}{h},$$

$$\Delta_y = \frac{\phi(x, y+h) - \phi(x, y)}{h}.$$

These differences involve five points around (x, y) (see Figure 3.1.6). Now we shall develop a random walk realization of (3.1.41), which we write out explicitly below:

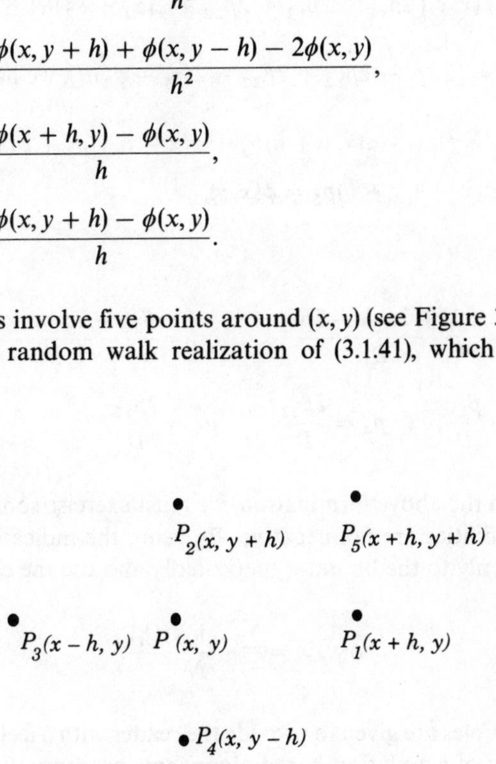

$P_2(x, y+h)$ $P_5(x+h, y+h)$

$P_3(x-h, y)$ $P(x, y)$ $P_1(x+h, y)$

$P_4(x, y-h)$

Figure 3.1.6

$$\begin{aligned}(3.1.43)\quad &\beta_{11}\frac{\phi(x+h,y)+\phi(x-h,y)-2\phi(x,y)}{h^2}\\ &+2\beta_{12}\frac{\phi(x+h,y+h)-\phi(x,y+h)-\phi(x+h,y)+\phi(x,y)}{h^2}\\ &+\beta_{22}\frac{\phi(x,y+h)+\phi(x,y-h)-2\phi(x,y)}{h^2}\\ &+2\alpha_1\frac{\phi(x+h,y)-\phi(x,y)}{h}+2\alpha_2\frac{\phi(x,y+h)-\phi(x,y)}{h}=0.\end{aligned}$$

Regrouping the terms in (3.1.43) gives

$$\begin{aligned}(3.1.44)\quad &\phi(x+h,y)(\beta_{11}+2\alpha_1 h-2\beta_{12})+\phi(x,y+h)(\beta_{22}+2\alpha_2 h-2\beta_{12})\\ &+\phi(x-h,y)\beta_{11}+\phi(x,y-h)\beta_{22}+\phi(x+h,y+h)2\beta_{12}\\ &=\phi(x,y)[2\beta_{11}-2\beta_{12}+2\beta_{22}+2(\alpha_1+\alpha_2)h].\end{aligned}$$

Now letting $D=[2\beta_{11}-2\beta_{12}+2\beta_{22}+2(\alpha_1+\alpha_2)h]$, we have

$$\begin{aligned}(3.1.45)\quad &\phi(x+h,y)p_1+\phi(x,y+h)p_2+\phi(x-h,y)p_3+\phi(x,y-h)p_4\\ &+\phi(x+h,y+h)p_5=\phi(x,y),\end{aligned}$$

where

$$(3.1.46)\quad p_1=\frac{\beta_{11}+2\alpha_1 h-2\beta_{12}}{D};\qquad p_2=\frac{\beta_{22}+2\alpha_2 h-2\beta_{12}}{D};$$

$$p_3=\frac{\beta_{11}}{D};\qquad p_4=\frac{2\beta_{22}}{D};\qquad p_5=\frac{2\beta_{12}}{D}.$$

We note that in the above formulation, we must exercise some care to ensure that the probabilities are nonnegative. By using the indicated probabilities, we walk randomly to the boundary repeatedly and use the estimate

$$(3.1.47)\qquad \phi(x,y)=\frac{\sum_{i=1}^{N}n_i f(Q_i)}{N}.$$

The above examples are given to provide the reader with a feel for the practical implementation of simulation-based algorithms as alternatives to the usual numerical approximation techniques. A certain amount of practice quickly brings the user to a point where he or she can write simulation algorithms in days to problems that would require the numerical analyst months to approach. In a later section, we shall give an example of such a case study.

Other advantages of the simulation approach could be given. For example, since walks to boundaries are so simple in execution, it is easy to conceptualize the utilization of parallel processors to speed up the computations with a minimum of handshaking between the CPUs.

But the main advantage of simulation is its ability to enable the user to bypass the traditional path in Figure 3.1.1 and go directly from the microaxioms to their macroconsequences. Our algorithm for solving the Fokker–Planck problem and our algorithm for solving the Dirichlet problem are not simply analogues of the "classical" differential equation formulations of these systems. *They are in fact descriptions of the axioms that typically give rise to these problems.* Here we note that the classical differential–integral equation formulation of many of the systems of physics and chemistry proceeds from the axioms that form the simulation algorithm. It was simply the case, in a precomputer age, that the differential–integral equation formulation appeared to give the best hope of approximate evaluation via series expansions and the like.

We now consider a class of problems where simulation is even more strongly suggested as the method of computational choice than for the above examples. Let us consider the problem of evaluating Wiener integrals of the form

$$\int_C F(x)\, d\mu_W(x), \tag{3.1.48}$$

which we define constructively as follows. C is the class of continuous functions on the interval $[0, T]$. Let $x(0) = 0$. Then divide the interval $[0, T]$ into k intervals of width T/k. Let $x(jT/k)$ be a normal random variable with mean $x(\{j-1\}T/k)$ and variance equal to $T/\{2k\}$. We shall start at $t = 0$ and generate $x(T/k)$ as an observation from $N(0, T/\{2k\})$. Then, stepping to $2T/k$, we generate $x(2T/k)$ from $N(X(T/k), T/\{2k\}$, and so on. Then we compute

$$I_1 = \frac{F(0) + F(X(T/k)) + F(X(2T/k)) + \cdots + F(X(T))}{k+1}. \tag{3.1.49}$$

After repeating (3.1.49) over N such paths, we compute

$$I(N, k) = \frac{1}{N}\sum_{j=1}^{N} I_j. \tag{3.1.50}$$

The limit as k and N go to infinity is the definition of (3.1.48). Now there are procedures for attempting to reformulate (3.1.48) in differential–integral equation form. But clearly, we have, from the very definition of a Wiener integral, the preferred computational and perceptual form.

In the coming years, simulation will tend to replace the venerable, but increasingly obsolete, closed form oriented approaches, which have shown themselves frequently to be obstacles rather than bridges to pass from the micro to the macro. In the case of such fields as econometrics, there may be general agreement by experts at the microaxiom level. But such is the complexity of obtaining the closed form solution from these axioms that the macro closed form is simply not available and so we are left with ad hoc, generally linear, models that have predicted 20 of the last three recessions. Here, in the social sciences, simulation frequently becomes not simply computationally and perceptually superior to classical approaches but the only realistic hope for obtaining useful representations of reality. The challenge for leadership in the development of this new simulation-oriented modeling and consequent software is one of the most promising and demanding ever to confront the statistics profession.

References

Lattes, Robert (1969). *Methods of Resolution for Selected Boundary Problems in Mathematical Physics,* Gordon & Breach, New York.

Shreider, Yuri A. (1966). *The Monte Carlo Method: The Method of Statistical Trials,* Pergamon Press, New York.

3.2. SIMDAT: AN ALGORITHM FOR DATA-BASED SIMULATION

There will be some situations in which we need to perform an analysis of what might happen on the basis of what has happened without constructing precise, well thought out models of the generating mechanisms involved. For example, we may have a complex network of modules which together generate an outcome. Suppose that for one or more of these modules we have extensive vector data bases. One such module might be a marketing survey involving consumer preferences and characteristics of the people surveyed. Even though we might not have the leisure or ability to construct a model that explains the results of the survey, we still wish to incorporate the results of our survey in a large-scale simulation of likely outcomes of the introduction of a new product. In this section, we shall develop a simple algorithm for this purpose (see Figure 3.2.1).

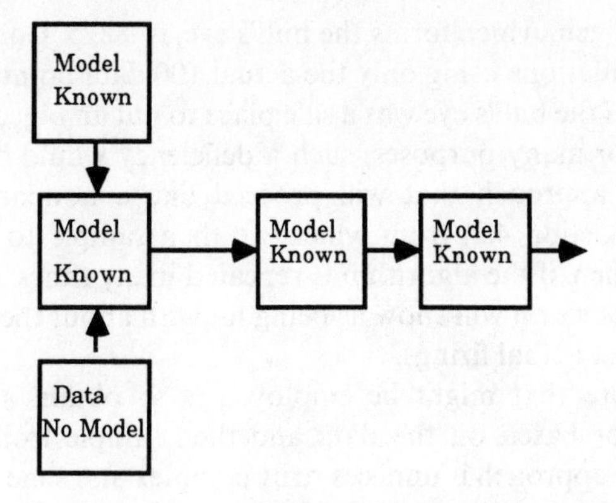

Figure 3.2.1

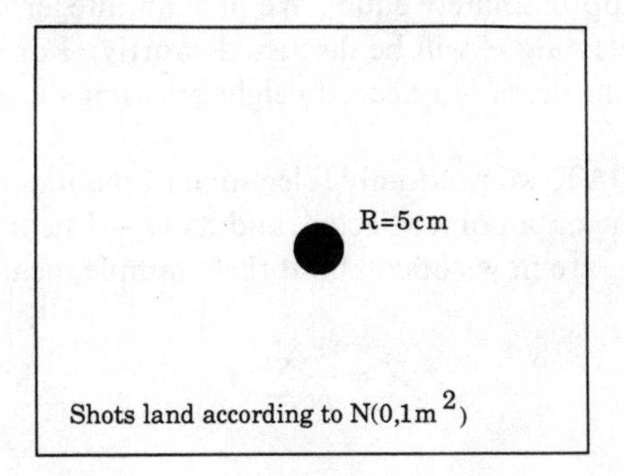

Figure 3.2.2

Let us suppose we have a random sample $\{X_j\}_{j=1 \text{ to } n}$ of k-dimensional vectors. We wish to generate pseudorandom vectors from the underlying, but unknown, distribution that gave rise to the random sample. We shall employ a Radical Pragmatist approach here, since we assume that the generating mechanism is too difficult to model readily. One approach we might employ in our simulation is simply to draw from the n data points.

A difficulty here is immediately demonstrated by an example. Suppose 100 bullets have been fired at a conventional bull's eye type target (Figure 3.2.2). We shall assume that the bullets strike the target with impact coordinates in the vertical and horizontal directions (actually, but unknown to us) as independent normal variates with mean zero and standard deviation 1 m (the center of the bull's eye being taken as (0, 0)). It can easily be determined (see Section A.1) that the probability that none of the 100 bullets will land in a circle of radius 5 cm, which forms the bull's eye, is .8825. Consequently, if we created our simulations using only the actual 100 data points, we could infer that the center of the bull's eye was a safe place to put an object of 5 cm radius.

Obviously, for many purposes, such a deficiency would be unacceptable. We require an approach that will proceed like a nonparametric density estimator (see Section 4.2) from which we then sample to obtain pseudo-observations. Then, if the algorithm is repeated many times, an object that is in a dangerous position will show as being hit with about the same frequency as would occur in actual firings.

One procedure that might be employed is to obtain a nonparametric density estimator based on the data and then sample from the estimated density. Such an approach is unnecessarily complex and time consuming. The solution we employ here is much easier.

We first carry out a rough rescaling, so that the variability in each of the k dimensions is approximately equal. We pick an integer m between 1 and n (the method of selecting m will be discussed shortly). For each of the n data points, we determine the $m - 1$ nearest neighbors using the ordinary Euclidean metric.

To start SIMDAT, we randomly select one of the n data points. We then have m vectors, the data point selected, and its $m - 1$ nearest neighbors. The vectors $\{X_j\}_{j=1 \text{ to } m}$ are now coded about their sample mean

$$\bar{X} = \frac{1}{m}\sum_{i=1}^{m} X_i, \tag{3.2.1}$$

to yield

$$\{X_j'\} = \{X_j - \bar{X}\}_{j=1}^{m}. \tag{3.2.2}$$

Next, we generate a random sample of size m from the one-dimensional uniform distribution

$$U\left(\frac{1}{m} - \sqrt{\frac{3(m-1)}{m^2}}, \frac{1}{m} + \sqrt{\frac{3(m-1)}{m^2}}\right). \tag{3.2.3}$$

This particular uniform distribution is selected to provide the desired moment properties below. Now the linear combination

$$X' = \sum_{l=1}^{m} u_l X_l' \tag{3.2.4}$$

is formed, where $\{u_l\}_{l=1 \text{ to } m}$ is a random sample from the uniform distribution in (3.2.3). Finally, the translation

$$X = X' + \bar{X} \tag{3.2.5}$$

restores the relative magnitude, and X is a simulated vector which we propose to be representative of the multivariate distribution that generated the original data set. To obtain the next simulated vector, we randomly select another point from the original data base and repeat the above sequence (sampling with replacement).

Now to motivate the algorithm, we consider briefly the sampled vector X_1 and its $m - 1$ nearest neighbors:

$$\begin{bmatrix} x_{1l} \\ x_{2l} \\ \vdots \\ x_{kl} \end{bmatrix}_{l=1 \text{ to } m} = \{X_l\}_{l=1}^m. \tag{3.2.6}$$

Let us suppose that this collection of observations represents a truncated distribution with mean vector μ and covariance matrix Σ. Let $\{u_l\}_{l=1 \text{ to } m}$ be an independent random sample from the uniform distribution (3.2.3). Then,

$$E(u_l) = \frac{1}{m}; \qquad \mathrm{Var}(u_l) = \frac{m-1}{m^2}; \qquad \mathrm{Cov}(u_i, u_j) = 0, \quad \text{for } i \neq j. \tag{3.2.7}$$

Forming the linear combination

$$Z = \sum_{l=1}^{m} u_l X_l, \tag{3.2.8}$$

we have, for the rth component of vector Z, $z_r = u_1 x_{r1} + \cdots + u_m x_{rm}$, the following relations:

$$E(z_r) = \mu_r, \tag{3.29}$$

$$\mathrm{Var}(z_r) = \sigma_r^2 + \frac{m-1}{m}\mu_r^2, \tag{3.2.10}$$

and

$$\mathrm{Cov}(z_r, z_s) = \sigma_{rs} + \frac{m-1}{m}\mu_r \mu_s. \tag{3.2.11}$$

We observe that the above construction provides Z's that are uncorrelated. Now if the mean vector of X were $\mu = (0, 0, \ldots, 0)'$, then the mean vector and covariance matrix of Z would be identical to those of X; that is, $E(z_r) = 0$, $\text{Var}(z_r) = \sigma_r^2$, and $\text{Cov}(z_r, z_s) = \sigma_{rs}$. In the less idealized situation with which we are confronted, the translation to the sample mean of the nearest neighbor cloud should result in the simulated observation having very nearly the same mean and covariance structure as that of the (truncated) distribution of the points in the nearest neighbor cloud. This has been substantiated empirically in many actual cases.

Now for m moderately large, by the central limit theorem, SIMDAT essentially samples from n Gaussian distributions with the mean and covariance matrices corresponding to those of the n, m nearest neighbor clouds. Accordingly, we obtain, in practice, results similar to those which would have been obtained by using a Gaussian kernel density estimator (see Section 4.2) with locally estimated covariance matrix with weight $1/n$ at each of the data points as the density from which pseudo-observations are to be drawn. The algorithm simply exploits the fact that our desired end product is a pseudo-sample rather than a density estimator to achieve computational and perceptual efficiencies.

The selection of m is not particularly critical. Naturally, if we let $m = 1$, we are simply sampling from the data set itself (this is Efron's "bootstrap"), and the difficulties in the bull's eye example may confront us. When we use too large a fraction of the total data set, we tend to obscure fine detail. But the selection of m is not the crucial matter that it is in the area of nonparametric density estimation. Let us consider below a mixture of three bivariate normal distributions. We shall generate 85 data points and 85 pseudodata points using two values for m. The means and covariance matrices of each are given below.

$$\mu_1 = \begin{bmatrix} -1 \\ -2 \end{bmatrix}; \quad \mu_2 = \begin{bmatrix} -2 \\ 3 \end{bmatrix}; \quad \mu_3 = \begin{bmatrix} 2 \\ \frac{3}{2} \end{bmatrix}. \tag{3.2.12}$$

$$\Sigma_1 = \begin{bmatrix} 1 & \frac{1}{2} \\ -\frac{1}{2} & 1 \end{bmatrix}; \quad \Sigma_2 = \begin{bmatrix} 1 & \frac{1}{2} \\ \frac{1}{2} & 1 \end{bmatrix}; \quad \Sigma_3 = \begin{bmatrix} 1 & \frac{1}{10} \\ \frac{1}{10} & 1 \end{bmatrix}. \tag{3.2.13}$$

The corresponding mixing proportions are $p_1 = \frac{1}{2}$, $p_2 = \frac{1}{3}$, $p_3 = \frac{1}{6}$, respectively. In Figure 3.2.3, where $m = 5$, we note a high degree of fidelity of the pseudodata to the actual data.

In Figure 3.2.4, we use a value of $m = 40$. Note that this value is actually half that of $n = 85$, the size of the data set. As expected, this oversmoothing tends to draw the pseudodata points more to the center. However, even for such an outlandish choice of m we note that the agreement between the data and the pseudodata is not bad for some purposes.

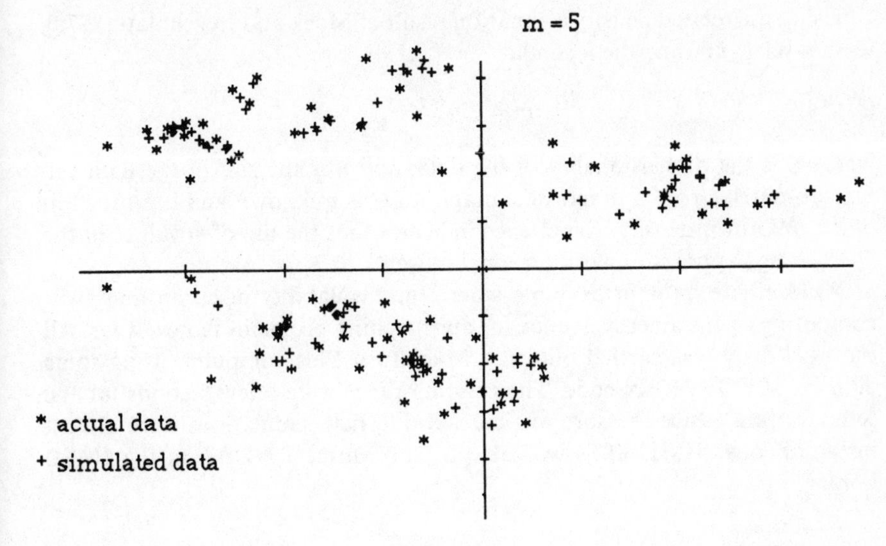

Figure 3.2.3

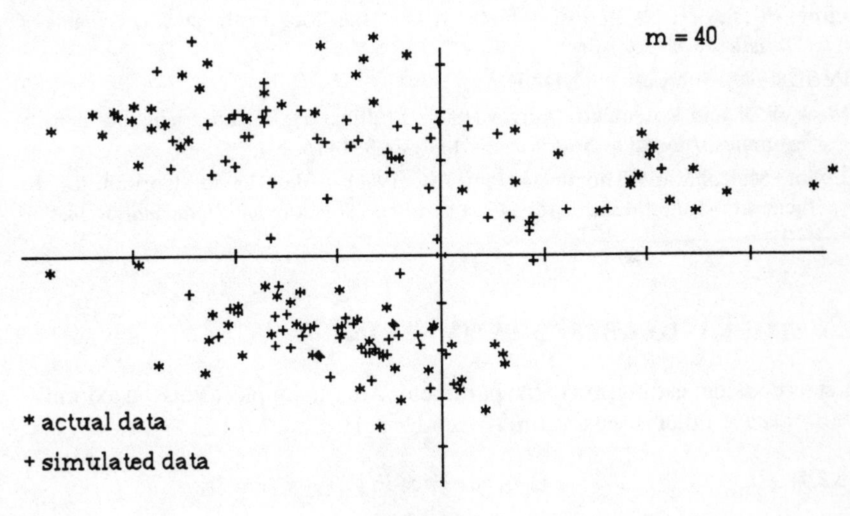

Figure 3.2.4

Using the related density estimation result of Mack and Rosenblatt (1979), we can select m using the formula

$$m = Cn^{4/(k+4)}, \tag{3.2.14}$$

where k is the dimensionality of the data, and n is the size of the data set. Unfortunately, from a practical standpoint, C is unknown and is a function of the unknown density. Experience indicates that the use of m values in the 1–5% range appears to work reasonably well.

We see here another example where the availability of cheap and swift computing opens an easy avenue for approaching problems in new ways. All the work here was carried out on a Macintosh Plus computer, using some 50 lines of FORTRAN code. The running times were a few seconds for the setup sorting. Once the sort was carried out, new simulations were almost instantaneous. SIMDAT is available as subroutine RNDAT in the IMSL library.

References

Efron, Bradley (1979). Bootstrap methods—another look at the jacknife, *Annals of Statistics,* **7,** 1–26.

IMSL (1989). Subroutine RNDAT.

Mack, Y. P. and Rosenblatt, Murray (1979). Multivariate k-nearest neighbor density estimates, *Journal of Multivariate Analysis,* **9,** 1–15.

Taylor, Malcolm and Thompson, James R. (1986). A data based algorithm for the generation of random vectors, *Computational Statistics and Data Analysis,* **4,** 93–101.

3.3. SIMULATION-BASED ESTIMATION

Let us consider estimation of the parameter λ in the simplest Poisson axiomitization (see a fuller discussion in Appendix A.1).

$$P(1 \text{ occurrence in } [t, t+\varepsilon]) = \lambda\varepsilon, \tag{3.3.1}$$

$$P(\text{more than 1 occurrence in } [t, t+\varepsilon]) = o(\varepsilon) \tag{3.3.2}$$

(where $\lim_{\varepsilon \to \infty} o(\varepsilon)/\varepsilon = 0$),

$$\begin{aligned} &P(k \text{ in } [t_1, t_2] \text{ and } m \text{ in } [t_3, t_4]) \\ &\quad = P(k \text{ in } [t_1, t_2])P(m \text{ in } [t_3, t_4]) \quad \text{if } [t_1, t_2] \cap [t_3, t_4] = 0, \end{aligned} \tag{3.3.3}$$

$$P(k \text{ in } [t_1, t_1 + s]) = P(k \text{ in } [t_2, t_2 + s]) \quad \text{for all } t_1, t_2, \text{ and } s. \tag{3.3.4}$$

As we show in Section A.1, the density of an occurrence given the simple Poisson axioms is simply

$$f(t|\lambda) = \lambda e^{-\lambda t} \quad \text{if } t \geq 0 \tag{3.3.5}$$
$$= 0, \quad \text{otherwise.}$$

Since we have the explicit form of the density readily available, the natural approach, on the basis of n failures $\{t_1, t_2, \ldots, t_n\}$, would be to use one of the several classical means of parametric estimation. For example, we might try maximum likelihood; that is, we seek the value of λ which maximizes

$$L(\lambda|t_1, t_2, \ldots, t_n) = \prod_{j=1}^{n} f(t_j|\lambda) = \lambda^n \exp\left(-\lambda \sum_{j=1}^{n} t_j\right). \tag{3.3.6}$$

This is readily accomplished by taking the logarithm of both sides of (3.3.6), differentiating with respect to λ, and setting the derivative equal to zero. This gives

$$\hat{\lambda} = \frac{n}{\sum_{j=1}^{n} t_j} = \frac{1}{\bar{t}}. \tag{3.3.7}$$

If, however, we did not have readily available the density of failure times, the classical methods of estimation would first of all require that we obtain this density. This might require a great deal of work. On the other hand, we might attempt a simulation strategy proceeding directly from the axioms that describe the process.

Using the sample $0 < t_1 < t_2 < \cdots < t_n$, we divide the time axis into k bins, the lth of which contains n_l observations. Assuming a value for λ, we use the simulation mechanism $SM(\lambda)$ to generate a large number N of simulated failures $0 < s_1 < s_2 < \cdots < s_N$. The number of these observations which fall into the lth bin will be denoted by v_{kl}. If our selection of λ was close to the truth, then the simulated bin probabilities

$$\hat{p}_{kl} = \frac{v_{kl}}{N} \tag{3.3.8}$$

should approximate the corresponding proportion of data in the same bin,

$$\hat{p}_l = \frac{n_l}{n}. \tag{3.3.9}$$

We shall call the asymptotic value of $\hat{p}_{kl}(\lambda)$ as N goes to infinity, $p_{kl}(\lambda)$. Criteria are needed for assessing the deviation of the $\hat{p}_{kl}(\lambda)$ from the $\hat{p}_l$. One criterion is the multinomial log likelihood

$$S_1(\lambda) = \ln(n!) - \sum_{j=1}^{k} \ln(n_j!) + \sum_{j=1}^{k} n_j \ln(\hat{p}_{kj}). \tag{3.3.10}$$

The simulated observations are used to estimate the probability $\hat{p}_{kj}$ that for the value of λ assumed an observation will fall in the jth bin. The expression shown is then the logarithm of the probability that for each j the jth bin will contain n_j observations. The first two terms in the expression for S_1 do not depend on λ but on the binned observations n_j. Consequently, we can drop them and use as the criterion the equivalent expression (now using the vector parameter Θ)

$$S_1(\Theta) = \sum_{j=1}^{k} n_j \ln(\hat{p}_{kj}). \tag{3.3.11}$$

This logarithm of the multinomial likelihood is maximized when

$$\hat{p}_{kj} = \frac{n_j}{n} = \hat{p}_j, \tag{3.3.12}$$

that is, when the simulated cell probabilities match those from the original sample.

The determination of the relative sizes of the k bins used to discretize the data has not been specified. There are a number of reasons for using a binning scheme with equal numbers of observations in each bin. For example, this minimizes the chance of empty cells in the simulation. Setting $p_j = 1/k$ gives at truth the Min Max Var$\{\hat{p}_{kj}\}$. Moreover, equal binning (setting $\hat{p}_j(\Theta) = 1/k$) guarantees that the expectation of $S_1(\Theta)$ does not increase for small perturbations of Θ from truth.

However, there are circumstances in which equal sized binning is not practical. For example, the values of the dependent variable may be clustered because of inaccuracies in measurement or rounding (e.g., when pathologists measure the diameters of tumors, values of 1 cm tend to be much more likely than, say, values of 1.142 cm). In this case, the estimation procedure is enhanced if the division points for discretizing the data are chosen between clusters.

Now expanding $\ln(\hat{p}_{ki})$ in a Taylor's series about p_{ki}, we have, discarding terms of $O(1/N^2)$, the following formula for the asymptotic variance of S_1:

$$\text{(3.3.13)} \quad \operatorname{Var}(S_1(\Theta)) = \sum_{i=1}^{k}\sum_{j=1}^{k} \frac{\partial S_1}{\partial \hat{p}_{ki}}\bigg|_{p_{ki}} \frac{\partial S_1}{\partial \hat{p}_{kj}}\bigg|_{p_{kj}} \operatorname{Cov}(\hat{p}_{ki}, \hat{p}_{kj})|_{p_{ki}p_{kj}}$$

$$- \sum_{i=1}^{k} \frac{(1 - p_{ki})}{Np_{ki}} n_i^2 - 2 \sum_{i=1}^{k-1} \sum_{j=i+1}^{k} \frac{p_{ki}p_{kj}}{Np_{ki}p_{kj}} n_i n_j$$

$$= \frac{1}{N}\left(\sum_{i=1}^{k} \frac{(1 - p_{ki})}{p_{ki}} n_i^2 - (n^2 - n_1^2 - \cdots - n_k^2)\right).$$

Clearly, (3.1.13) is minimized for $p_{ki} = n_i/n$ for all i.

Let us now suppose for the remainder of this section that the bins are chosen so that $\hat{p}_i = n_i/n = 1/k$ for all i. We shall assume that the parameter being estimated is more general than the scalar λ and shall denote it by the vector $\Theta = (\theta_1, \theta_2, \ldots, \theta_m)$. Then if Θ is close to the truth, the simulated bin probabilities should each approximate $1/k$. If, for a given k, there is only one value of Θ such that, for $l = 1, 2, \ldots, k$,

$$\text{(3.3.14)} \qquad \lim \hat{p}_{kl} = \frac{1}{k}, \quad \text{as } n \to \infty \text{ and } N \to \infty,$$

then we say that Θ is *k-identifiable.*

For the simplest Poisson axiomitization (given in (3.3.1)–(3.3.4)), the bin boundaries $b_{n0}, b_{n1}, b_{n2}, \ldots, b_{nk}$ converge almost surely to

$$\text{(3.3.15)} \qquad b_l = \frac{-\ln(1 - (l/k))}{\lambda_0},$$

where λ_0 is the true value of λ. Suppose there is only one bin ($b_0 = 0$ and $b_1 = \infty$). Now, for any value of λ, all the simulated failures will fall into the bin. Consequently, in this case, λ is not 1-identifiable. For two bins, $b_0 = 0$, $b_2 = \infty$ and

$$\text{(3.3.16)} \qquad b_1 = -\frac{\ln(\frac{1}{2})}{\lambda_0}.$$

There is no value of λ other than λ_0 for which

$$\text{(3.3.17)} \qquad \lim_{N \to \infty} \hat{p}_{kl}(\lambda) = \tfrac{1}{2}.$$

Consequently, for the simple Poisson distribution, λ is 2-identifiable. Moreover, for values of $k \geq 2$, λ is *k-identifiable.* If $SM(\Theta)$ is k-identifiable, then a natural procedure for estimating Θ is to pick the value that maximizes

$$S_2(\Theta) = \sum_{l=1}^{k} \ln(\hat{p}_{kl}(\Theta)). \tag{3.3.18}$$

Note that maximizing S_2 is equivalent to maximizing S_1 when all the n_j are equal.

Typically, the size of the sample, n, will be relatively small compared to N, the size of the simulation. It is clear that the number of bins, k, is a natural smoothing parameter. For example, for $k = 1$, $\mathrm{Var}(\hat{p}_{11}) = 0$ for all n and N. The variability of $S_2(\Theta)$ can be approximated via the asymptotic formula

$$\begin{aligned}\mathrm{Var}\, S_2(\Theta) &= \sum_{i=1}^{k}\sum_{j=1}^{k} \frac{\partial S_2}{\partial \hat{p}_{ki}}\bigg|_{p_{ki}} \frac{\partial S_2}{\partial \hat{p}_{kj}}\bigg|_{p_{kj}} \mathrm{Cov}(\hat{p}_{ki}, p_{kj})|_{p_{ki}p_{kj}} \\ &= \sum_{i=1}^{k} \frac{1}{p_{ki}^2(\Theta)} \frac{p_{ki}(\Theta)(1 - p_{ki}(\Theta))}{N} \\ &\quad - 2\sum_{i=1}^{k-1}\sum_{j=i+1}^{k} \frac{p_{ki}(\Theta)p_{kj}(\Theta)}{Np_{ki}(\Theta)p_{kj}(\Theta)}.\end{aligned} \tag{3.3.19}$$

Clearly, then

$$\mathrm{Var}(S_2(\Theta)) \approx \frac{1}{N}\left(\sum_{i=1}^{k} \frac{1 - \hat{p}_{ki}}{\hat{p}_{ki}} - k(k-1)\right). \tag{3.3.20}$$

Now,

$$E(S_2(\Theta)) \approx \sum_{i=1}^{k} \ln(\hat{p}_{ki}). \tag{3.3.21}$$

Thus, we have a ready measure of a *signal-to-noise ratio* via

$$SN_2(k) = \frac{|E(S_2(\Theta))|}{\sqrt{\mathrm{Var}(S_2(\Theta))}} \approx \frac{\left|\sqrt{N}\sum_{i=1}^{k} \ln(\hat{p}_{ki})\right|}{\sqrt{\sum_{i=1}^{k} \frac{1 - \hat{p}_{ki}}{\hat{p}_{ki}} - k(k-1)}}. \tag{3.3.22}$$

Let us suppose that

$$\frac{\mathrm{Max}\{p_{kl}\}}{\mathrm{Min}\{p_{kl}\}} = M. \tag{3.3.23}$$

Suppose that l of the bins have probability η and $k - l$ of the bins have probability $M\eta$. Then, asymptotically,

$$\text{Var}(S_2(\Theta)) = \frac{1}{N}\left(\frac{(M-1)l + k}{M\eta} - k^2\right). \tag{3.3.24}$$

This variance is maximized for $l = k/2$. So for the "worst case,"

$$\text{Var}(S_2(\Theta)) = \frac{k^2}{N}\left(\frac{(M+1)^2}{4M} - 1\right). \tag{3.3.25}$$

Moreover,

$$E(S_2(\Theta)) = \frac{k}{2}\ln M - k\ln\left(\frac{(M+1)k}{2}\right). \tag{3.3.26}$$

Thus

$$SN_2(k, M) = \frac{\sqrt{N}\,|\ln(2\sqrt{M}/k(M+1))|}{\sqrt{((M+1)^2/4M) - 1}}. \tag{3.3.27}$$

In Table 3.3.1 (from Thompson et al (1987)) we show values of $SN_2(k, M)/\sqrt{N}$ for various k and M. For example, for $M = 100$ and 20 bins, a simulation size of 11,562 would be required to achieve a signal-to-noise ratio of 100. For $M = 10$, a simulation size of 1014 will achieve a signal-to-noise ratio of 100. At any stage of the optimization algorithm, we can use $\text{Max}\{\hat{p}_{kl}\}/\text{Min}\{\hat{p}_{kl}\}$ as a pessimistic estimate of M to achieve conservative estimates of the signal-to-noise ratio.

In a practical situation we shall be confronted with situations where we have two values of Θ—say, Θ_1 and Θ_2—and wish to know whether

Table 3.3.1. Values of $SN_2(k, M)/\sqrt{N}$

	M			
k	5	10	50	100
5	2.13	1.52	0.83	0.65
10	2.90	2.00	1.03	0.79
20	3.68	3.14	1.23	0.93
100	5.48	3.63	1.70	1.26

$$\lim_{N_1 \to \infty} S_2(\Theta_1) > \lim_{N_2 \to \infty} S_2(\Theta_2). \tag{3.3.28}$$

Now suppose for a particular pair of simulated sample sizes we do have

$$S_2(\Theta_1) > S_2(\Theta_2). \tag{3.3.29}$$

How do we know whether or not this difference is significant? From (3.3.20) we can obtain $\mathrm{Var}(S_2(\Theta_1))$ and $\mathrm{Var}(S_2(\Theta_2))$. The variance of the difference is given approximately by

$$\mathrm{Var}(S_2(\Theta_1) - S_2(\Theta_2)) = \mathrm{Var}(S_2(\Theta_1)) + \mathrm{Var}(S_2(\Theta_2)). \tag{3.3.30}$$

Thus, if

$$S_2(\Theta_1) - S_2(\Theta_2) > 2\sqrt{\mathrm{Var}(S_2(\Theta_1) - S_2(\Theta_2))}, \tag{3.3.31}$$

we are reasonably confident that the difference is real and not due to simulation noise.

As an example, let us suppose we have two values of Θ—Θ_1 and Θ_2. We have divided the sample of n failures into 10 bins and carried out simulations with sizes $N_1 = 900$ and $N_2 = 2500$. Suppose, moreover, the estimated cell probabilities are as shown in Table 3.3.2 from Thompson et al (1987).

From (3.3.20) and (3.3.30) we have

$$\mathrm{Var}(S_2(\Theta_1) - S_2(\Theta_2)) = \frac{3.5786}{900} + \frac{14.304}{2500} = 0.009678, \tag{3.3.32}$$

Table 3.3.2. Estimated Bin Probabilities

	Θ_1	Θ_2
$\hat{p}_{10,1}$	.12	.09
$\hat{p}_{10,2}$	.08	.11
$\hat{p}_{10,3}$	.10	.05
$\hat{p}_{10,4}$	.12	.08
$\hat{p}_{10,5}$	.08	.09
$\hat{p}_{10,6}$	.09	.05
$\hat{p}_{10,7}$	.11	.15
$\hat{p}_{10,8}$	.07	.11
$\hat{p}_{10,9}$	.12	.12
$\hat{p}_{10,10}$	.11	.15

giving

$$2\sqrt{\text{Var(diff)}} = 0.196955. \tag{3.3.33}$$

Moreover, from (3.3.19), $S_2(\Theta_1) = -23.1966$ and $S_2(\Theta_2) = -26.6579$. Since $S_2(\Theta_1) > S_2(\Theta_2) + 2\sqrt{\text{Var(diff)}}$, we can be reasonably confident that the apparently preferred performance of Θ_1 is not simply due to simulation noise. If, however, the difference between $S_2(\Theta_1)$ and $S_2(\Theta_2)$ is not significant, we may increase the two simulation sizes to increase the signal-to-noise ratio. We note that if any cell is empty, $S_2(\Theta) = -\infty$ and is essentially not informative. Accordingly, we need a procedure to avoid using a mesh structure that is too fine, particularly at the beginning of the iteration procedure when we may be far from the optimum. One such procedure is to examine the $\{\hat{p}_{kj}\}$ for a choice of k (say, 100). We then find the largest $\hat{p}_{kj}$, say, M. Then, starting at the leftmost bin if $\hat{p}_{kj} < M/100$, we combine bin j with bins to the right until the combined bins have total $\hat{p}_{k,j} + \hat{p}_{k,j+1} + \hat{p}_{k,j+2} + \cdots + \hat{p}_{k,j+l} \geq M/100$. When this has been achieved, we replace each of the $\hat{p}_{k,j}, \hat{p}_{k,j+1}, \hat{p}_{k,j+2}, \ldots, \hat{p}_{k,j+l}$ with

$$(\hat{p}_{k,j} + \hat{p}_{k,j+1} + \hat{p}_{k,j+2} + \cdots + \hat{p}_{k,j+l})/(l+1).$$

Another convenient criterion function is Pearson's *goodness of fit*

$$S_3(\Theta) = \sum_{j=1}^{k} \frac{(\hat{p}_{kj} - \hat{p}_j)^2}{\hat{p}_j}. \tag{3.3.34}$$

Obviously, this function is minimized when $\hat{p}_{kj} = \hat{p}_j$ for all j.

The criterion function S_3 has an advantage over both S_1 and S_2. Suppose both $\hat{p}_{k1} = c\hat{p}_1$ and $\hat{p}_{k2} = c\hat{p}_2$. Then S_3 is unchanged when the two cells are combined into a single cell.

Next, we observe that the variability of each of the three criterion functions considered is nondecreasing in the number of cells, k (in the completely noninformative case when $k = 1$, each of the three criterion functions has zero variability). On the other hand, increasing k increases our ability to discriminate between the effectiveness of various Θ's to produce simulations that mimic the behavior of the actual sample of failure times.

To demonstrate this fact, let us suppose that we consider the effect of combining the first two cells when S_3 is used. Let us suppose the "miss" of $\hat{p}_{k1} + \hat{p}_{k2}$ from $\hat{p}_1 + \hat{p}_2$ is an amount η. Then for the pooled sample, the contribution to S_3 is

$$\frac{\eta^2}{\hat{p}_1 + \hat{p}_2}. \tag{3.3.35}$$

Now for these cells uncombined, let

$$\hat{p}_{k1} = \hat{p}_1 + \frac{\hat{p}_1}{\hat{p}_1 + \hat{p}_2}\eta + \varepsilon\eta; \tag{3.3.36}$$

$$\hat{p}_{k2} = \hat{p}_2 + \frac{\hat{p}_2}{\hat{p}_1 + \hat{p}_2}\eta - \varepsilon\eta.$$

In the uncombined case, the contribution to S_3 is

$$\frac{[\hat{p}_1/(\hat{p}_1 + \hat{p}_2) + \varepsilon]^2\eta^2}{\hat{p}_1} + \frac{[\hat{p}_2/(\hat{p}_1 + \hat{p}_2) - \varepsilon]^2\eta^2}{\hat{p}_2} \tag{3.3.37}$$

$$= 1 + \frac{\varepsilon^2(\hat{p}_1 + \hat{p}_2)^2/\hat{p}_1\hat{p}_2}{\hat{p}_1 + \hat{p}_2}\eta^2 \geq \frac{\eta^2}{\hat{p}_1 + \hat{p}_2}.$$

We note that only in the case where η is split between the two cells in proportion to $\hat{p}_1$ and $\hat{p}_2$ does a decrease in the number of cells fail to decrease S_3. Hence, a decrease in the number of cells decreases our ability to tell us how well a simulation is mimicking the actual data. A similar argument holds for S_1 and S_2.

Naturally, a number of cells greater than the size of the actual sample would be a bad idea. As practical matter, using the exhaustive statistic $(0 < t_1 < t_2 < \cdots < t_n)$ to give the cell boundaries $(0, t_1], (t_1, t_2], \ldots, (t_n, \infty)$ would generally be extreme. We recall that our strategy is to select a value of Θ that gives a simulation mimicking the sample. But, in a broader sense, we seek to mimic samples that *could have happened*. For n large, $F(t \leq t_j)$ will be very nearly j/n except for j nearly 0 or for j close to n. For these values of j, the t_j are poor estimators for the (j/n)tiles of $F(\cdot)$. Thus, for the leftmost and rightmost bins, we might be well advised to see to it that at least 1% of the sample observations are included in each.

We now address the issue of a practical means of obtaining a 95% confidence set for the true value of Θ. Once the algorithm has converged to a value, say, Θ^*, we then use this value to generate M simulated data sets, each of size n. We then determine

$$\overline{S_j} = \frac{1}{M}\sum_{i=1}^{M} S_j(\Theta^*, T_i) \tag{3.3.38}$$

and

$$s_{S_j}^2 = \frac{1}{M}\sum_{i=1}^{M} [S_j(\Theta^*, T_i) - \overline{S_j}]^2, \tag{3.3.39}$$

where T_i denotes the ith simulated data set of size n and T_0 is the actual sample. Then, with roughly 95% certainty,

$$S_j(\Theta) = \bar{S}_j \pm \frac{2}{\sqrt{M}} s_{S_j}. \tag{3.3.40}$$

Next, using Θ^* as the center around which we may pick additional values of Θ to enable the fitting of a quadratic curve, we have

$$S_j(\Theta) = A + B\Theta + C\Theta^T\Theta. \tag{3.3.41}$$

The 95% confidence set for Θ can now be approximated by

$$S_j(\Theta^*, T_0) - \frac{2}{\sqrt{M}} s_{S_j} \leq A + B\Theta + C\Theta^T\Theta \leq S_j(\Theta^*, T_0) + \frac{2}{\sqrt{M}} s_{S_j}. \tag{3.3.42}$$

Let us return to an example from Section 1.5 dealing with adjuvant chemotherapy. We recall that the approximate formula suggested (1.5.9) for an adjuvant cure was a rather serious underestimate for small values of adjuvant cure and that the exact formula (1.5.16) was bothersome to derive. The reason, once again, for the difficulty in derivation of the exact formula has to do with the fact that when we are seeking a particular end result, for example, no resistant metastases, we must essentially seek out all those paths that could have led to the particular end result. Thus, we are constrained to work backwards, against the forward time path of the Poissonian axiomitization. For this, we generally pay a heavy, frequently an unbearable, penalty.

Happily, it is very frequently the case that we can use a forward path to deal with our optimization problems if we are willing to replace our reliance on a "closed form" with a simulation-based strategy. Let us indicate such a strategy to deal with the adjuvant chemotherapy formulation. We recall that the probability of no metastases by the time a tumor reached a size N was given by

$$P(\text{no metastases by } N) = e^{-\mu(N-1)}. \tag{3.3.43}$$

We can immediately write down the distribution function of the occurrence of metastases:

$$F(t) = 1 - e^{-\mu(n-1)}. \tag{3.3.44}$$

Moreover, if we wish to do so, we can use $n(t)$ rather t as our "clock." Thus, we may write

$$F(n) = 1 - e^{-\mu(n-1)}. \tag{3.3.45}$$

Furthermore, if the last metastasis has occurred when the total tumor mass is n_j, then we have for the distribution function of the primary tumor volume at the time of the next metastatic event

$$F(n_{j+1}) = 1 - e^{-\mu(n_{j+1}-n_j)}. \tag{3.3.46}$$

Since t is a continuous random variable, we know that F is itself distributed as a $U(0, 1)$ random variable. Thus, we have a ready means of simulating the times of metastatic events. If the last metastasis occurred at a tumor size of n_j, then the size of the tumor mass at the time of the next metastatic formation can be simulated by generating u from the uniform distribution on the unit interval and then computing

$$n_{j+1} = n_j - \frac{1}{\mu}\ln(1 - u). \tag{3.3.47}$$

Once a metastasis has formed (from one cell), we assume that it grows at the same rate as the primary. The probability that such a metastatic tumor, which appears when the primary tumor is of size n, will not develop resistance before the primary tumor reaches size N is seen to be given by

$$P(\text{no resistance develops before primary has size } N) = e^{-\beta((N/n)-1)}. \tag{3.3.48}$$

Equations (3.3.47) and (3.3.48) give us an immediate and direct means for simulating the probability that a patient will not first develop a metastasis which subsequently develops resistance as a function of tumor size N. We give an appropriate flowchart in Figure 3.3.1.

A similar flowchart can be used to find the probability that a patient will not have developed, by the time the primary tumor size is N, first a resistant clone in the primary which then has metastasized away from the primary. We note that the simulation-based approach also enables a much greater degree of generalization than "closed form" attempts. For example, we can easily enable the possibility that each new clone has growth characteristics that are different from those of the primary. At the time of formation of a new clone, we can sample from an appropriate density of growth rates to be used for the new clones.

Next, let us consider a model for cancer progression. Suppose we have a set of data (time from discovery of primary to discovery of second tumor $\{t_1, t_2, \ldots, t_n\}$ from patients who, having had a primary tumor removed, subsequent to the surgery, exhibited a second cancer in another part of the

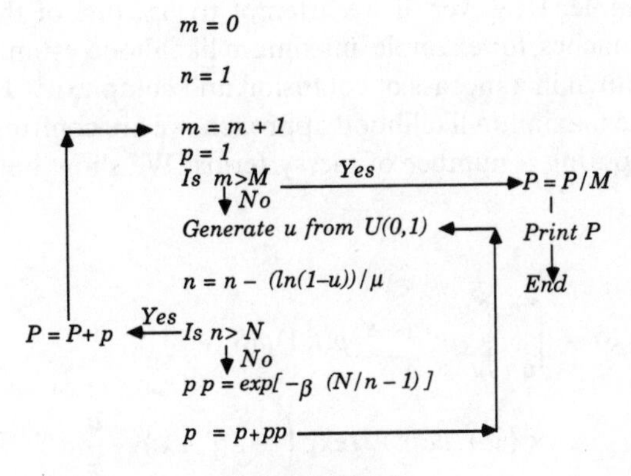

Figure 3.3.1

body. The classical interpretation of such a secondary cancer is that it is, as discussed above, the result of a cell breaking off from the primary tumor, migrating to another part of the body, and continuing to grow exponentially. However, it has been postulated by Bartoszyński et al. (1982) that the second tumor may be a metastasis *or* it may be simply a second cancer that has no descent from the primary but is the result of the continuing susceptibility of the patient to the formation of a particular kind of cancer. We wish to develop a method using the data at hand to determine the relative importance of the metastatic and systemic processes.

As usual, we use Occam's razor to give a testable set of the simplest postulates consistent with our conjecture. These would seem to be the following:

(3.3.49a) For each patient, each tumor originates from a single cell and grows exponentially at rate α.

(3.3.49b) The probability that a tumor of size $Y_j(t)$, not previously detected and removed prior to time t, is detectable in $[t, t + \Delta t]$ is $bY_j(t)\Delta t + o(\Delta t)$.

(3.3.49c) Until the removal of the primary, the probability of metastasis in $[t, t + \Delta t]$ is $aY_0(t)$, where $Y_0(t)$ is the mass of the primary tumor.

(3.3.49d) The probability of systemic occurrence of a tumor in $[t, t + \Delta t]$ is $\lambda \Delta t + o(\Delta t)$ independent of the prior history of the patient.

Written as they are, in standard Poissonian forward form, the postulates are extremely simple. However, if we attempt to use one of the backward closed form approaches, for example, maximum likelihood estimation, we are quickly bogged down in a morass of confusion and complexity. For example, in order to use the maximum likelihood approach, we are confronted with the necessity of computing a number of messy terms. We show one of these in (3.3.50).

(3.3.50)

$$P(T_1 = S', T_2 > S) = \int_0^\infty \int_u^\infty e^{W(S-S')} p(t; 1) p(S'; e^{\alpha u})$$
$$\times (\lambda + a e^{\alpha(t-u)}) \exp\left(-\lambda(t-u) - \frac{a}{\alpha}(e^{\alpha(t-u)} - 1) \right)$$
$$\times H(v(S - S'); S'; e^{\alpha S'}) H(v(S - S') e^{\alpha S'}; u; e^{\alpha(t-u)})\, du\, dt$$
$$+ \int_0^\infty \int_u^\infty e^{W(S-S')} p(t; 1) \exp\left(-\lambda t \frac{a}{\alpha}(e^{\alpha t} - 1) \right) \lambda e^{-\lambda u}$$
$$\times p(S' - u; 1) H(v(S - S'); S' - u; 1)\, du\, dt,$$

where

(3.3.51)
$$H(s; t; z) = \exp\left(\frac{az}{\alpha} e^{\alpha t}(e^s - 1) \log[1 + e^{-s}(e^{-\alpha t} - 1)] + \frac{\lambda s}{\alpha} - \frac{\lambda}{\alpha} \log[1 + e^{\alpha t}(e^s - 1)] \right),$$

(3.3.52)
$$p(t; z) = bze^{\alpha t} \exp\left(-\frac{bz}{\alpha}(e^{\alpha t} - 1) \right),$$

(3.3.53)
$$w(y) = \lambda \left(\int_0^y e^{-v(u)}\, du - y \right),$$

and $v(u)$ is determined from

$$u = \int_0^v \frac{ds}{a + b + \alpha s - ae^{-s}}. \tag{3.3.54}$$

The order of computational complexity here is roughly that of four-dimensional quadrature. This is near the practical limit of contemporary mainframe computers. The time required for the estimation of the four parameters in the above model was roughly 2 hours using Chandler's (1969) robust optimization routine STEPIT on the CYBER 173. The running time using the Nelder–Mead optimization algorithm described in Section A.2. is approximately the same.

Excessive running times are by no means our only difficulty using a classical backward approach, such as maximum likelihood. It is extremely arduous to keep track of all the ways that an event might have occurred. If we leave one out, our computation may be worthless. Then there is the problem of human error. When we note an absurd numerical output from our program, we must face the possibility that it could be due to a model that simply does not conform to the data, or it could be due to an error in flowcharting, or it could be due to a programming error, or it could be due to roundoff error, or it could be due to a dozen other possible flaws. Going back and checking each is troublesome, time consuming, and psychologically draining. Furthermore, if we need to modify the model, it will frequently be necessary to scrap our work and begin again if we use the classical backward approach. In the putting together of the formulas and computations in Bartoszyński et al. (1982), approximately 1.5 person years were required. It was this hitting the wall both in terms of computer time and person time that led to the conception of SIMEST.

To use a simulation-based approach, we are only required to find a way of simulating times of discovery of secondary tumors as a function of the parameters (α, a, b, λ). To accomplish this task, we make the following definitions:

(3.3.55a) t_D = time of detection of primary tumor;

(3.3.55b) t_M = time of origin of first metastasis;

(3.3.55c) t_S = time of origin of first systemic tumor;

(3.3.55d) t_R = time of origin of first recurrent tumor;

(3.3.55e) t_d = time from t_R to detection of first recurrent tumor;

(3.3.55f) t_{DR} = time from t_D to detection of first recurrent tumor.

We generate all random variables by first generating u from a uniform distribution on the interval $[0, 1]$. Then we set $t = F^{-1}(u)$, where F is the appropriate cumulative distribution function. Now, assuming the tumor volume at time t is $v(t) = ce^{\alpha t}$, where c is the volume of one cell, it follows from (3.3.49) that

$$F_D(t_D) = 1 - \exp\left(-\int_0^{t_D} bce^{\alpha\tau}\,d\tau\right) = 1 - \exp\left(-\frac{bc}{\alpha}e^{\alpha t_D}\right). \tag{3.3.55}$$

Similarly, we find

$$F_M(t) = 1 - \exp\left(-\frac{ac}{\alpha}e^{\alpha t_M}\right), \tag{3.3.56}$$

$$F_S(t) = 1 - e^{-\lambda t_S}, \tag{3.3.57}$$

and

$$F_d(t_d) = 1 - \exp\left(-\frac{bc}{\alpha}e^{\alpha t_d}\right). \tag{3.3.58}$$

We can now write our simulation algorithm quite easily (Figure 3.3.2).

Using the actual sample $t_1, t_2, \ldots, t_n$, we can generate k bins, each with apparent probabilities $p_1, p_2, \ldots, p_k$. Now we can use the simulation algorithm, $SM1(\alpha, \lambda, a, b)$, N times to generate a simulation of N recurrences $s_1, s_2, \ldots, s_n$. The numbers of simulated detections in each of these bins will be denoted by $v_{k1}, v_{k2}, \ldots, v_{kk}$. The simulated bin probabilities are then computed by

```
Input α, λ, a, b
Repeat until a > 0
Generate t_D
Generate t_M
If t_M > t_D, then t_M ← ∞
Generate t_S
t_R ← min{t_M, t_S}
Generate t_d
t_DR ← t_R + t_d − t_D
s = t_DR
If s < 0, discard
End repeat
Return s
```

Figure 3.3.2. $SM1(\alpha, \lambda, a, b)$.

(3.3.59) $$\hat{p}_{kj} = \frac{v_{kj}}{N}.$$

Then, using $S_1(\alpha, \lambda, a, b)$, $S_2(\alpha, \lambda, a, b)$, $S_3(\alpha, \lambda, a, b)$, or some other reasonable criterion, we are in a position to ascertain how well our guessed value of (α, λ, a, b) mimics the behavior of the sample. The results obtained agreed with the closed form approach utilized by Bartoszyński et al. (1982). Furthermore, the program was debugged and running satisfactorily after 2 person days of work rather than 1.5 person years of work following the classical approach.

We note that our $SM1$ algorithm embodies a simplifying assumption utilized in the closed form maximum likelihood approach. This assumption was necessary in the closed form approach, since the complexity of a backward approach would have been too great without it. Using the forward simulation-based approach, this assumption can easily be relaxed as we show in $SM2$.

In Figure 3.3.3, we generate an array of metastasis detection times $\{t_{aM}(j)\}$ and systemic detection times $\{t_{as}(j)\}$ and pick the smallest of these as the first detection time of a secondary tumor. The forward simulation approach enables us to increase the complexity of the underlying model at modest marginal cost to the simulation algorithm.

Since the simulation algorithm is forward, modularization of the routines is easy and natural; new modules may be added on as desired. For example, suppose we wished to add a fifth postulate.

```
Generate t_D
j = 0
i = 0
Repeat until t_M(j) > t_D
j = j + 1
Generate t_M(j)
Generate t_dM(j)
t_dM(j) ← t_dM(j) + t_M(j)
If t_dM(j) < t_D, then t_dM(j) ← ∞
Repeat until t_S > 10t_D
i = i + 1
Generate t_dS(i)
t_dS(i) ← t_dS(i) + t_S(i)
s ← min{t_dM(j), t_dS(i)}
Return s
End Repeat
```

Figure 3.3.3. $SM2(\alpha, \lambda, a, b)$.

Generate u from $U(0,1)$
If $u > \gamma$, then proceed as in $SM2$
If $u < \gamma$, then proceed as in $SM2$ except replace the step
"Repeat until $t_S(i) > 10t_D$" with the step "Repeat until $t_S(i) > t_D$"

Figure 3.3.4. $SM3(\alpha, \lambda, a, b, \gamma)$.

(3.3.49e) A fraction γ of the patients ceases to be at systemic risk at the time of removal of the primary tumor if no secondary tumors exist at that time. A fraction $1 - \gamma$ of the patients remain at systemic risk throughout their lives.

We can then simulate the modified (3.3.49) axioms via $SM3$.

An endless array of other modifications to the axiomatic system can be made at little cost to simulation complexity. As stated, a principal argument in favor of the simulation approach as opposed to the classical backward closed form approach is that the forward direction in which we flowchart the simulation is much easier than the backward approach used uniformly by the classical algorithms. Perhaps surprisingly, our experience is that computing time is also much less for the simulation approach proposed. For example, for the metastatic versus systemic scenario considered above, the running time for the simulation-based estimation approach was roughly 10% of that for the classical approach.

A recent approach of Atkinson, Brown, and Thompson (1988) gives a means of implementation which is useful in higher-dimensional situations where Cartesian binning becomes unwieldy. We start out with a real-world system observable through k-dimensional observations X. We believe that the generating system can be approximately described by a model characterized by the parameter Θ. If we have a data set from the system of size n, we can, for a value of Θ, simulate a quasidata set of size N. Then, we compute the sample mean vector and covariance matrix of the X data set. Next, we transform the data set to a transformed set $U = AX + b$ with mean zero and identity matrix I. We then apply the transformation to the simulated data set. We compute the sample mean $\bar{X}$ and covariance matrix Σ of the simulation data set. If the simulated set is essentially the same as the actual data set, then it should have for its transformed values mean zero and identity covariance matrix I. If the underlying data distribution is not too bizarre, we can measure the fidelity of the simulation data to the actual data by computing the ratio of the Gaussian likelihoods of the transformed simulated data sets using the mean and covariance estimates from the actual data and the simulated data, respectively. Defining

(3.3.60) $$Q(u_{1i}, u_{2i}, \ldots, u_{ki}) = \sum_{j,l=1}^{k} \sigma^{jl}(u_{ji} - \bar{X}_j)(u_{li} - \bar{X}_l),$$

where σ^{jl} is the j, lth element of the inverse of Σ, we have

(3.3.61) $$L(\Theta) = \frac{\prod_{i=1}^{N} \frac{1}{(2\pi)^{k/2}} \exp\left[-\frac{1}{2}(u_{1i}^2 + u_{2i}^2 + \cdots + u_{ki}^2)\right]}{\sum_{i=1}^{N} \frac{\sqrt{|\Sigma^{-1}|}}{(2\pi)^{k/2}} \exp\left[-\frac{1}{2}Q(u_{1i}, u_{2i}, \ldots, u_{ki})\right]}.$$

This approach has proved quite effective even in some cases where data sets are not unimodal. For more complex structures, we transform the data as above but then, for each point in the actual data set, find the distance of the point d_D to its nearest neighbor in the data set and the distance d_S to its nearest neighbor in the simulated data set (taking $N = n - 1$). We then use as a measure of fidelity of the simulated data to the actual data

(3.3.61) $$Q(\Theta) = \sum_{i=1}^{n} \frac{(d_{Di} - d_{Si})^2}{d_{Di}}.$$

References

Atkinson, E. Neely, Bartoszyński, Robert, Brown, Barry W., and Thompson, James R. (1983). Simulation techniques for parameter estimation in tumor related stochastic processes, in *Proceedings of the 1983 Computer Simulation Conference*, North Holland, New York, pp. 754–757.

Atkinson, E. Neely, Bartoszyński, Robert, Brown, Barry W., and Thompson, James R. (1983). Maximum likelihood techniques, in *Proceedings of the 44th Meeting of the International Statistical Institute, Contributed Papers*, **2**, 494–497.

Atkinson, E. Neely, Brown, Barry W., and Thompson, James R. (1988). SIMEST and SIMDAT: Differences and convergences, *Proceedings of the 20th Symposium on the Interface: Computing Science and Statistics*, to appear.

Bartoszyński, Robert, Brown, Barry W., McBride, Charles, and Thompson, James R. (1981). Some nonparametric techniques for estimating the intensity function of a cancer related nonstationary Poisson process, *Annals of Statistics*, **9**, 1050–1060.

Bartoszyński, Robert, Brown, Barry W., and Thompson, James R. (1982). Metastatic and systemic factors in neoplastic progression, in *Probability Models and Cancer*, Lucien LeCam and Jerzy Neyman, eds., North Holland, New York, pp. 253–264.

Chandler, J. P. (1969). STEPIT, *Behavioral Science*, **14**, 81.

Nelder, J. A. and Mead, R. (1965). A simplex method for function minimization. *Computational Journal*, **7**, 308–313.

Thompson, James R., Atkinson, E. Neely, and Brown, Barry W. (1987). SIMEST: An algorithm for simulation-based estimation of parameters characterizing a stochastic process, in *Cancer Modeling,* James R. Thompson and Barry W. Brown, eds., Marcel Dekker, New York, pp. 387–415.

CHAPTER FOUR

Some Techniques of Nonstandard Data Analysis

4.1. A GLIMPSE AT EXPLORATORY DATA ANALYSIS

Books have been written on John W. Tukey's revolutionary technique of exploratory data analysis (which is generally referred to simply as EDA), and I can only hope in a brief discussion to shed some light on the fundamentals of that subject. Moreover, the point of view that I take in this section represents my own perceptions, which may be very different from those of others. Some of the enthusiasts of EDA frequently take a philosophical position which I would characterize as being very strongly toward that of the Radical Pragmatist position in the Introduction. A common phrase that one hears is that "EDA allows the data to speak to us in unfettered fashion." The "fetters" here refer to preconceived models that can get between us and the useful information in the data. The position might be characterized by Will Rogers' famous dictum, "It isn't so much ignorance which harms us. It's the things we know that aren't so."

Whereas I believe that perceptions are always in the light of preconceived models, which we hope to modify and see evolve, there is much more to EDA than the antimodel position of some of its adherents. It is this "much more" about which I wish to speak. The digital computer is a mighty device in most quantitative work these days. Yet it has serious limitations that did not so much apply to the now discarded analog devices of the 1950s. Analog devices were very much oriented toward holistic display of the output of a model. They were not oriented toward dealing with mountains of data, nor were they particularly accurate. Digital devices, on the other hand, can be made as accurate as we wish and handle the storage and manipulation of digitized information extremely well.

At this point in time, we have hardware that is very much more "trees" oriented than "forest" oriented. We can easily ask that this or that set of operations be performed on this or that megabyte of encoded data. But we are increasingly aware of the cognitive unfriendliness of coping with digitally processed information. Analog devices were much closer to the way the human brain reasons than are digital devices.

Perhaps what is needed is a hybridized device that combines the strong points of both analog and digital computers. But such a hardware device will be years in bringing to a successful construction. In the meantime, what do we do? One approach might be simply to try to beat problems to death on the number cruncher. But such an approach quickly stalls. We have the computer power to obtain pointwise estimates of 10-dimensional density functions using data sets of sizes in the tens of thousands. But where shall we evaluate such a density function? How shall the computer be trained to distill vast bodies of information into summaries that are useful to us? These are difficult problems and the answers will be coming in for some time.

In the meantime, we need to cope. It is this necessity somehow to address the fact that the digital computer has outstripped our abilities to use the information it gives us that EDA addresses. Needing a good analog processor to handle the digital information and having none, a human observer is used to fulfill the analog function.

One recurring theme in science fiction has been the human who is plugged into a computer system. But the observer in EDA, unlike the sci-fi cyborg, is not hardwired into the system, is not deprived of freewill, is in fact in control of the digital system. One present limitation of exploratory data analysis is the slow input–output performance of freewilled human observers. Thus, man-in-the-loop EDA could not be used, for example, to differentiate between incoming missiles and decoys in the event of a large-scale attack. EDA is exploratory not only in the sense that we can use it for analyzing data sets with which we have little experience. We should also view EDA as an alpha step toward the construction of the analog–digital hybrid computer, which will not have the slow input–output speeds of the human–digital prototype.

In the discussion below, we shall address some of the important human perception bases of EDA. Let us give a short list of some of these:

1. The only function that can be identified by the human eye is the straight line.
2. The eye expects adjacent pixels to be likely parts of a common whole.
3. As points move far apart, the human processor needs training to decide when points are no longer to be considered part of the common whole. Because of the ubiquity of situations where the central limit theorem, in one form or another, applies, a natural benchmark is the normal distribution.

4. A point remains a point in any dimension.
5. Symmetry reduces the complexity of data.
6. Symmetry essentially demands unimodality.

Let us address the EDA means of utilizing the ability of the human eye to recognize a straight line. We might suppose that since linear relationships are not all that ubiquitous, the fact that we can recognize straight lines is not particularly useful. Happily, one can frequently reduce monotone relationships to straight lines through transformations. Suppose, for example, that the relationship between the dependent variable y and the independent variable x is given by

$$y = 3e^{0.2x} \tag{4.1.1}$$

We show a graph of this relationship in Figure 4.1.1.

We can easily see that the relationship between x and y is not linear. Furthermore, we see that y is increasing in x at a faster than linear rate. Further than this, our visual perceptions are not of great use in identifying the functional relationship.

But suppose that we decided to plot the logarithm of y against x as shown in Figure 4.1.2.

Now we have transformed the relationship between x and y to a linear one. By recalling how we transformed the data, we can complete our task of identifying the functional relationship between x and y. So, we recall that we

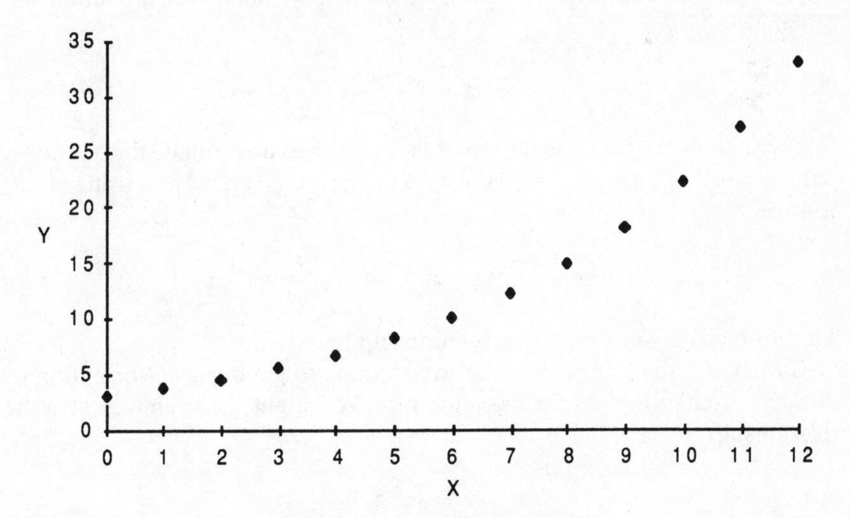

Figure 4.1.1. Untransformed data.

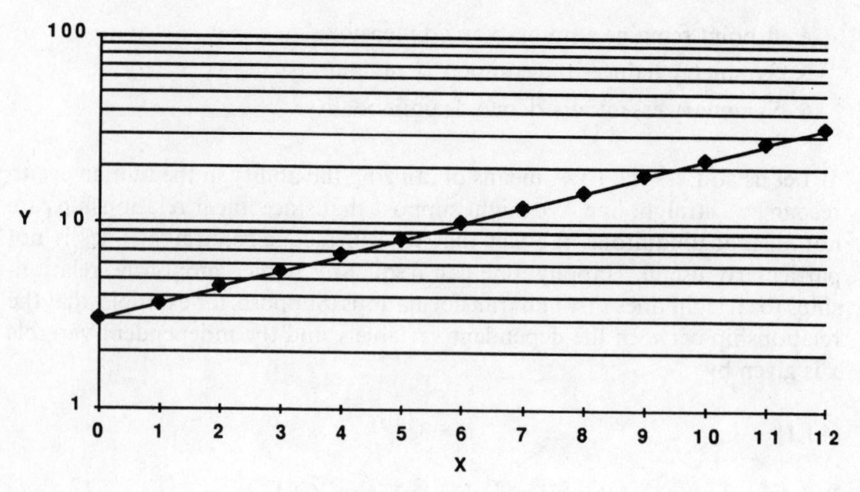

Figure 4.1.2. Transformed data.

started with an unknown functional relationship

$$y = f(x). \tag{4.1.2}$$

But then we saw that $\ln(y)$ was of the form

$$\ln(y) = a + bx. \tag{4.1.3}$$

Exponentiating both sides of (4.1.3), we see that we must have a relationship of the form:

$$y = e^a e^{bx}. \tag{4.1.4}$$

Once we know the functional form of the curve, we can estimate the unknown parameters by putting in two data pairs (x_1, y_1) and (x_2, y_2) and using (4.1.3) to solve

$$\ln(y_1) = a + bx_1 \quad \text{and} \quad \ln(y_2) = a + bx_2. \tag{4.1.5}$$

This immediately gives the true relationship in (4.1.1).

Clearly, we shall not always be so fortunate to get our transformation to linearity after trying simply a semilog plot. We might, for example, have the relationship

$$y = 3x^{0.4}. \tag{4.1.6}$$

In such a case, simply taking the logarithm of y will not give a linear plot, for

$$\ln(y) = \ln(3) + 0.4\ln(x) \tag{4.1.7}$$

is not linear in x. But, as we see immediately from (4.1.7), we would get a straight line if we plotted $\ln(y)$, not versus x, but versus $\ln(x)$. And, again, as soon as the transformation to linearity has been achieved, we can immediately infer the functional relationship between x and y and compute the parameters from the linear relationship between $\ln(y)$ and $\ln(x)$.

Now it is clear from the above that simply using semilog and log–log plots will enable us to infer functional relationships of the forms

$$y = ae^{bx} \tag{4.1.8}$$

and

$$y = ax^b, \tag{4.1.9}$$

respectively.

This technique of transforming to essential linearity has been used in chemical engineering for a century in the empirical modeling of complex systems in mechanics and thermodynamics. Indeed, the very existence of log–log and semilog graph paper is motivated by applications in these fields. In the classical applications, x and y would typically be complicated dimensionless "factors," that is, products and quotients of parameters and variables (the products and quotients having been empirically arrived at by "dimensional analysis") which one would plot from experimental data using various kinds of graph paper until linear or nearly linear relationships had been observed. But the transformational ladder of Tukey goes far beyond this early methodology by ordering the transformations one might be expected to use and approaching the problem of transformation to linearity in methodical fashion. For example, let us consider the shapes of curves in Figure 4.1.3. Now it is clear that curve A is growing faster than linearly. Accordingly, if we wish to investigate transformations that will bring its rate of growth to that of a straight line, we need to use transformations that will reduce its rate of growth. Some likely candidates in increasing order of severity of reduction are

$$\begin{aligned} &y^{1/2}, \\ &y^{1/4}, \\ &\ln(y), \\ &\ln(\ln(y)), \end{aligned} \tag{4.1.10}$$

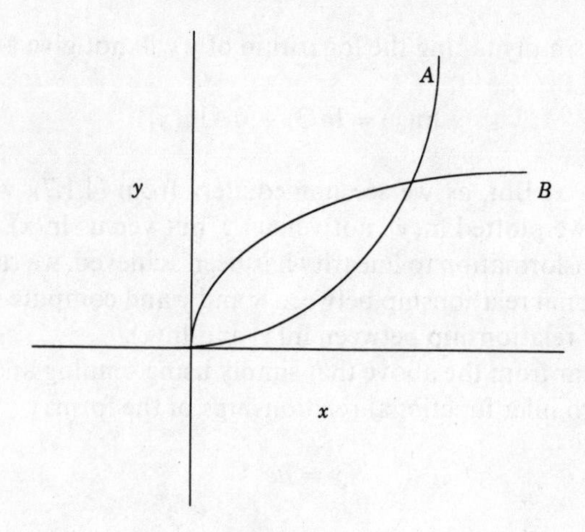

Figure 4.1.3

Similarly, if curve B is to be transformed to linearity, we might try, in decreasing order of severity,

(4.1.11)
$$\exp(e^y),$$
$$\exp(y),$$
$$y^4,$$
$$y^2.$$

Putting the two groups of transformations together, we can build a transformational ladder:

(4.1.12)
$$\exp(e^y)$$
$$\exp(y)$$
$$y^4$$
$$y^2$$
$$y$$
$$y^{1/2}$$
$$y^{1/4}$$
$$\ln(y)$$
$$\ln(\ln(y)).$$

The shape of the original y curve points us up or down the transformational ladder. Using the transformational ladder to find more complicated functional relationships between y and x becomes much more difficult. For example, it would require a fair amount of trial and error to infer a relationship such as

$$y = 4 + 2x^2 + x^3. \tag{4.1.13}$$

Furthermore, we must face the fact that in practice our data will be contaminated by noise. Thus, uniqueness of a solution will likely evade us.

For a great many situations, the use of Tukey's transformational ladder will bring us quickly to an understanding of what is going on. The technique avoids the use of a criterion function and uses the visual perceptions of an observer to decide the driving mechanism.

For more complicated problems, we can still be guided by the philosophy of the technique to use the computer to handle situations like that in (4.1.13) even when there is a good degree of noise contamination. We might decide, for example, to use least squares to go through a complex hierarchy of possible models, fitting the parameters as we went. So, we might employ

$$S(\text{model}(\text{in } x)) = \sum (y - \text{model})^2. \tag{4.1.14}$$

If we have an appropriately chosen hierarchy of models, we might have the computer output those that seemed most promising for further investigation. The problem of choosing the hierarchy is a nontrivial problem in artificial intelligence. We must remember, for example, that if models in the hierarchy are overparameterized, we may come up with rather bizarre and artificial suggestions. For example, if we have 20 data points, a 19th degree polynomial will give us a zero value for the sum in (4.1.14).

Let us now turn to the second of the perception-based notions of EDA: namely, the fact that the eye expects continuity, that adjacent points should be similar. This notion has been used with good effect, for example, in "cleaning up" NASA photographs. For example, let us suppose we have a noisy monochromatic two-dimensional photograph with light intensities measured on a Cartesian grid as shown in Figure 4.1.4. We might decide to smooth the

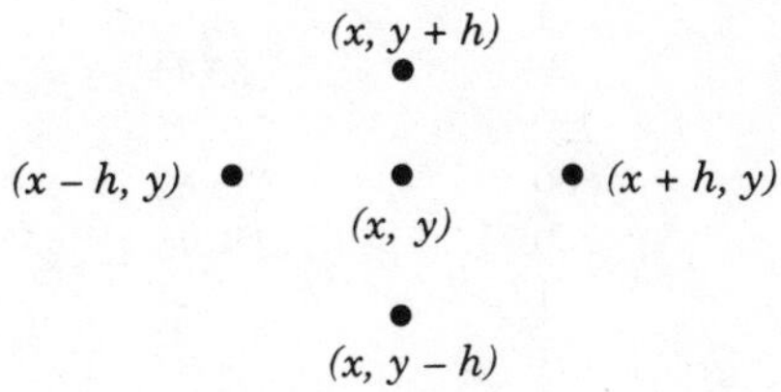

$(x, y + h)$

$(x - h, y)$ • • • $(x + h, y)$

(x, y)

•

$(x, y - h)$

Figure 4.1.4

intensities, via the hanning formula

(4.1.15)

$$I(x, y) \leftarrow \frac{4I(x, y) + I(x - h, y) + I(x, y - h) + I(x, y + h) + I(x + h, y)}{8},$$

where $I(x, y)$ is the light intensity at grid point (x, y).

Valuable though such a smoothing device has proved itself to be (note that this kind of device was used by Tukey and his associates 40 years ago in time series applications), there is the problem that outliers (wild points) can contaminate large portions of a data set if the digital filter is applied repeatedly. For example, suppose we consider a one-dimensional data set, which we smooth using the hanning rule:

$$I(x) \leftarrow \frac{2I(x) + I(x - h) + I(x + h)}{4}. \tag{4.1.16}$$

At the ends of the data set, we simply use the average of the end point with that of the second point. We show in Table 4.1.1 the data set followed by successive hanning smooths. We note that the wild value of 1000 has effectively contaminated the entire data set. To resolve this anomaly, Tukey uses a smooth based on medians of groups of three down the data set; that is, we use the rule

$$I(x) \leftarrow \text{Med}(I(x - h), I(x), I(x + h)). \tag{4.1.17}$$

The end points will simply be left unsmoothed in our discussion, although better rules are readily devised. In the data set above, the smoothing by threes

Table 4.1.1. Repeated Hanning Smooths

Data	H	HH	HHH
1	1	1	1
1	1	1	16.61
1	1	63.44	94.66
1	250.75	250.75	235.14
1000	500.50	375.62	313.18
1	250.75	250.75	235.14
1	1	63.44	94.66
1	1	1	16.61
1	1	1	1

approach gives us what one would presumably wish, namely, a column of ones.

As a practical matter, Tukey's median filter is readily used by the computer. It is a very localized filter, so that typically if we apply it until no further changes occur (this is called the 3R smoother), we do not spread values of points throughout the data set. Note that this is not the case with the hanning filter. Repeated applications of the hanning filter will continue to change the values throughout the set until a straight line results. Consequently, it is frequently appropriate to use the 3R filter followed by one application of the hanning filter (H). The combined use of the 3RH filter generally gets rid of the wild points (3R), and the unnatural plateaus of the 3R are smoothed by the H. Far more elaborate schemes are, of course, possible. We could, if we believed that two wild points could occur in the same block of three points, simply use a 5R filter.

In Table 4.1.2 we perform a 3RH smooth on a data set of daily unit productions on an assembly line. Figure 4.1.5 quickly shows how the 3RH

Table 4.1.2. Various Smooths

Day	Production	3	3R	3RH
1	150	150		157.5
2	165	165		168.25
3	212	193		188
4	193	201		199
5	201	201		201
6	220	201		199.5
7	195	195		190.25
8	170	170		176.25
9	161	170		167.75
10	182	161		160.25
11	149	149		142.25
12	110	110		117.5
13	95	101		101.75
14	101	95		87.75
15	60	60		64.25
16	42	42		46.5
17	15	42		46.5
18	110	60		55.5
19	60	80	60	60
20	80	60		57.5
21	50	50		50
22	40	40		45

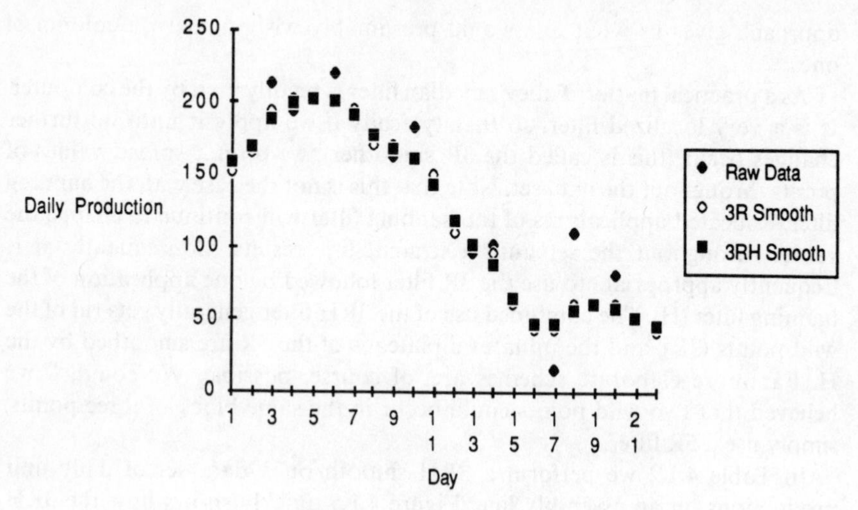

Figure 4.1.5. Raw and smooth.

smooth approximates closely what we would do if we smoothed the raw data by eye.

At this point, we should mention that all the smooths of EDA are curve fits, not derived models. We clearly find the 3RH smooth a more visually appealing graph than the raw data. But the data were measured precisely; the fluctuations really were there. So, in a sense, we have distorted reality by applying the 3RH smooth. Why have we applied it nevertheless? The human visual system tends to view and store in memory such a record holistically. Whether we smoothed the data or not, our eye would attempt to carry out more or less equivalent operations to those of 3RH. The human eye expects continuity and we do not readily perceive data digitally. The smooth gives us a benchmark (the forest) around which we can attempt to place the trees. For example, we might ask what was causing the unexpectedly low production on day 17. As we mentioned earlier, EDA tries to assist humans to carry out the analog part of the analysis process. The 3RH smooth done on the computer very nearly reproduces the processing carried out by the human eye. In a very real sense, Tukey's deceptively simple 3RH smooth is a powerful result in artificial intelligence.

Let us now address the third point, the making of the decision that a point has removed itself from a class by extreme behavior. We note that we have already addressed this point somewhat, since we have discussed the use of the median and hanning filters. If we seek a benchmark by which "togetherness" of a group of points can be measured, we might decide to use the ubiquitous normal distribution. We note that for this distribution,

(4.1.18) $$P[X < x] = \frac{1}{\sqrt{2\pi}} \int_{-\infty}^{z} e^{-t^2/2}\, dt,$$

where $z = (x - \mu)/\sigma$, with X having mean μ and standard deviation σ. For the normal distribution, the value $z = 0.675$ in (4.1.17) gives probability .75. By symmetry, the value $z = -0.675$ gives probability .25. Tukey calls the corresponding x values "hinges." The difference between these standardized values is 1.35. Let us call 1.5 times this H spread (interquartile range) a step. Adding a step to the standardized hinge gives a z value of 2.7. This value of 2.7 represents the standardized "upper inner fence." The probability a normal variate will be greater than the upper inner fence or less than the lower inner fence is $.007 \approx 1\%$. Adding another step to the upper inner fence gives the "upper outer fence" (in the standardized case with mean 0 and standard deviation 1, this will give $z = 4.725$). The probability of a normal variate not falling between the outer fences is .0000023, roughly two chances in a million. It could be argued that a value which falls outside the inner fences bears investigation to see whether it is really a member of the group. A value outside the outer fences is most likely not a member of the group. (Note that both these statements assume the data set is of modest size. If there are 1 million data points, all from the same normal distribution, we would expect 7000 to fall outside the inner fences and 2 to fall outside the outer fences.)

Let us examine a data set of annual incomes of a set of 30 tax returns supposedly chosen at random from those filed in 1938. Suppose the reported incomes are 700, 800, 1500, 2500, 3700, 3900, 5300, 5400, 5900, 6100, 6700, 6900, 7100, 7200, 7400, 7600, 7900, 8100, 8100, 8900, 9000, 9200, 9300, 9900, 10,400, 11,200, 13,000, 14,700, 15,100, and 16,900.

We first construct a "stem-and-leaf" plot with units in hundreds of dollars (Table 4.1.3). We note that the "plot" appears to be a hybrid between a table and a graph. In recording the actual values of the data, instead of only counts, Tukey's stem-and-leaf plot gives us the visual information of a histogram, while enabling full recovery of each data point. Here is an example where we can see both the forest and the trees.

From the stem-and-leaf plot, it is clear that certain tacit assumptions have been made. For example, we compute the "depths" from both ends of the set. Thus, a kind of symmetrical benchmark has been assumed. Let us further point to symmetry by computing the median (the average of the two incomes of depth 15 from the top and that of depth 15 from the bottom), namely, 7500 dollars. The two hinges can be obtained by going up to the two averages of incomes of depth 7 and 8. Thus, the lower hinge is 5350 and the upper hinge is 9600. A step is given by $(9600 - 5350)1.5 = 6375$. Thus, the two inner fences are given by -1025 and 15,975. The two outer fences are given by -7400 and 22,350. We note immediately one income (16,900) falls outside the inner fences, but none outside the outer fences.

Table 4.1.3. Stem-and-Leaf in Units of 100 Dollars

Depth		
2	0*	78
3	1	5
4	2	5
6	3	79
	4	
9	5	349
12	6	179
17	7	12469
13	8	119
10	9	0239
6	10	4
5	11	2
	12	
4	13	0
3	14	7
2	15	1
1	16	9

Let us now consider the various popular summary plots used for the income information. We have already seen one, the stem-and-leaf. Although this plot looks very much like a histogram turned on its side, we note that it shows not only the forest but also the trees, since we could completely recover our table from the plot. In the present situation, the stem-and-leaf might be sufficient data compression. Let us consider, however, some other plots.

The "five figure summary" plot in Figure 4.1.6 shows the mean, hinges, and extreme upper and lower incomes. Clearly, the five figure summary is much more compressed than the stem-and-leaf plot. But it draws emphasis to the supposed center of symmetry and looks at the hinges and extremal values. Naturally, as the sample becomes larger, we would expect that the median and the hinges do not change much. But the extremal values certainly will. A graphical enhancement of the five figure summary is the "box-and-whiskers" plot shown in Figure 4.1.7.

M15h	7500	
H7h	5350	9600
1	700	16900

Figure 4.1.6. Five figure summary.

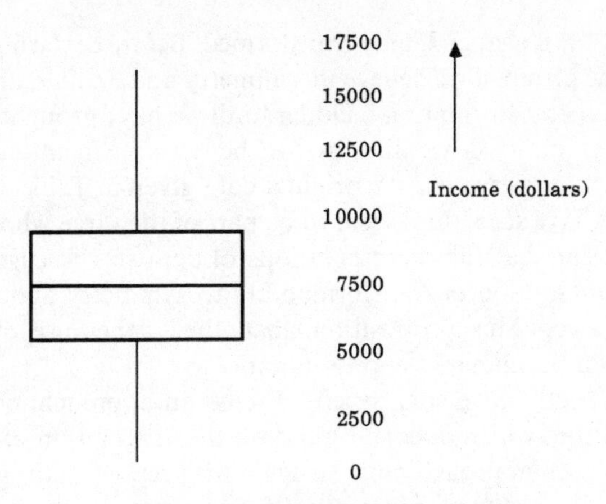

Figure 4.1.7. Box-and-whiskers plot.

A generally more useful plot than the box-and-whiskers representation is the "schematic plot" (Figure 4.1.8). Essentially, in this plot, the ends of the "whiskers" are the values inside the inner fences but closest to them. Such values are termed "adjacent." Essentially then, the schematic plot replaces the extremal values with the .0035 percentiles.

In Figure 4.1.8, we seem to have a data set that is not at all inconsistent with the assumption of being all "of a piece." We might have felt very differently if, say, we had been presented with the above income data which someone had mistakenly raised to the fourth power. Going through our standard analysis, we would find values outside the upper outer fence. Yet, the data have

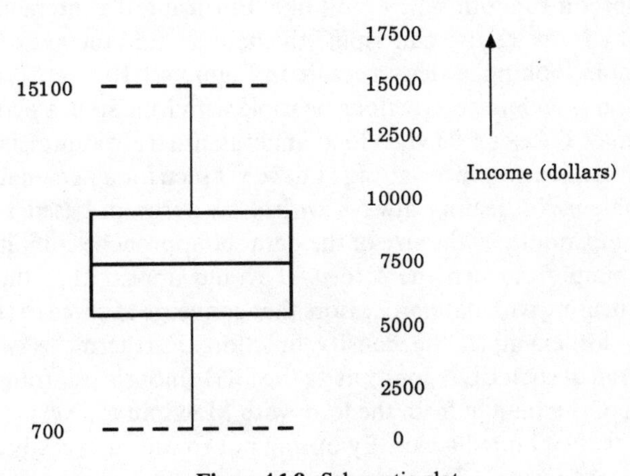

Figure 4.1.8. Schematic plot.

essentially not been changed, only transformed. Before declaring points to be untypical of the group, if we believe in symmetry and unimodality, we should run through our transformational ladder until we have brought the data to a state of near symmetry. If we did this, for the example mentioned, we would arrive at something very near the original data given in Table 4.1.3, and that data set, as we have seen, does seem to be part of the same whole.

Now it is clear that the representations of data sets discussed so far are built on the assumption of transformability to symmetry about an internal mode. If we accept this proposition, then the further use of the normal distribution as a benchmark is nontraumatic.

In the next section, we shall briefly discuss an approach, nonparametric density estimation, which does not build on the assumption of unimodality. Obviously, such an approach must struggle with representational difficulties about which EDA need not concern itself. There is a crucial issue here. How reasonable is it to assume unimodality and symmetry, and does this assumption get better or worse as the dimensionality of the data set increases? My own view is that the problem of dealing with the pathology of outliers (extremal points that are to be discarded from membership in the data set) is not as serious as that of multimodality, and that the even more serious problem of data lying in bizarre and twisted manifolds in higher-dimensional space ought to begin receiving more of our attention. One further issue that nonparametric density estimation investigators must face is that of representation of the density function suggested by the data. For higher-dimensional problems, EDA neatly sidesteps the representational issue by looking always at the original data points, rather than density contours. Let us consider two-dimensional projections of a three-dimensional data set generated by the routine RANDU. In Figure 4.1.9, we note what appears to be more or less what we expect a random set to look like. But using the interactive routine MacSpin (D^2 Software), we can "spin" the data around the axes, to arrive at the nonrandom looking lattice structure in Figure 4.1.10.

The human–machine interactions possible with MacSpin, a personal computer version of Tukey's PRIM-9 cloud analysis, are truly impressive, certainly the most impressive graphics package I have yet seen for a personal computer. Several problems of dealing always with a scattergram-based analysis are obvious. For example, as the size of the data set approaches infinity, the data points will simply blacken the screen. It would appear that there are advantages to dealing with data processors that converge to some fixed, informative entity—for example, the density function. Furthermore, whereas the automitization of such EDA concepts as the 3RH smooth are straightforward, the removal of the human from the loop with MacSpin is a very complicated problem in artificial intelligence. By opting not to use such easily automated concepts as contouring, EDA relies very much on the human eye to incorporate continuity in data analysis. Furthermore, by making assumptions of

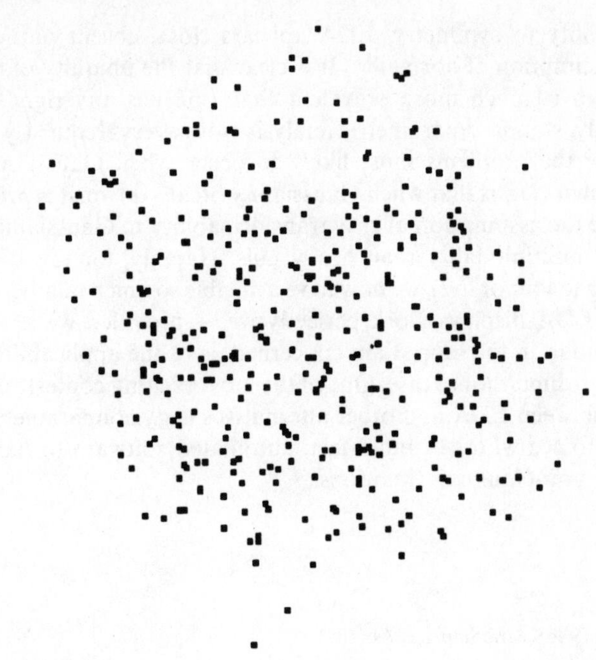

Figure 4.1.9

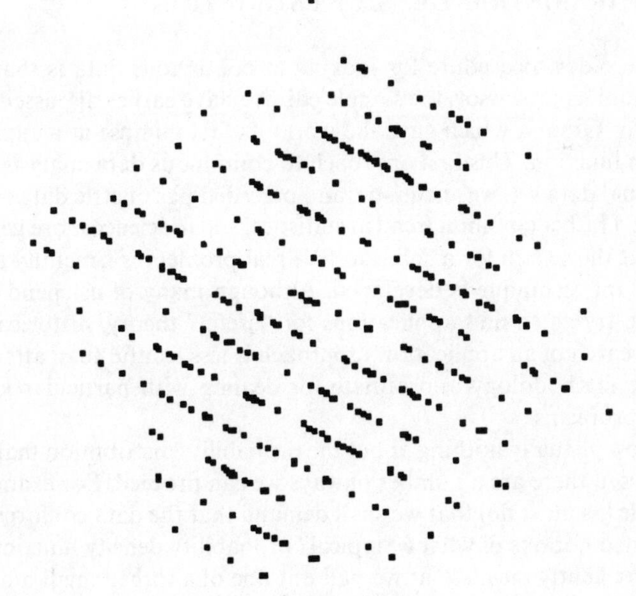

Figure 4.1.10

transformability to symmetry, EDA appears close, absent outliers, to the practical assumption of normality. It is clear that the ubiquity of the central limit theorem is much more prevalent than one has any right to expect. Consequently, standard parametric analysis works very frequently. The issue is: What are the problems most likely to occur when Gaussianity breaks down? My own view is that when Gaussianity breaks down, it is probably not safe to make the assumption of easy transformability to Gaussianity. I worry more about multimodality than heavy tails. Happily, for low-dimensional problems (up to four or five), we now have available so much number crunching power that EDA displays work perfectly well—provided we are happy to leave the human in the loop. I am concerned as to the applicability of EDA for the higher-dimensional case. One of the most exciting contests in statistics is the race between EDA and other alternatives (e.g., nonparametric density estimation) to see who can build fast, automated software to handle high-dimensional non-Gaussian situations.

References

D^2 Software (1986). *MacSpin* 1.1, Austin.

Tukey, John W. (1977). *Exploratory Data Analysis,* Addison-Wesley, Reading, MA.

4.2. NONPARAMETRIC DENSITY ESTIMATION

Perhaps the oldest procedure for looking at continuous data is that of the histogram and its precursor, the sample cdf. We have earlier discussed the life table of John Graunt, which gave the world its first glimpse at a cumulative distribution function. This first approach to continuous data analysis started with an actual data set, was heuristic, and preceded parametric data analysis. We see here a rather common trend in statistics, and in science more generally: namely, that the search for a solution to a real problem is generally the way that important technique is developed. Although many of us spend a great deal of time trying to find applications for "useful" theory, historically, the "theory in search of an application" approach is less fruitful than attempts to develop the methodology appropriate for dealing with particular kinds of real-world problems.

If we know virtually nothing about the probability distribution that generated a data set, there are a number of ways we can proceed. For example, we might decide (as most do) that we shall demand that the data conform to our predetermined notions of what a "typical" probability density function looks like. This frequently means that we pull out one of a rather small number of density functions in our memory banks and use the data to estimate the parameters characterizing that density. This is an approach that has been employed with varying degrees of success for a hundred years or so.

There is a strong bias in the minds of many toward the normal (also called Gaussian or Laplacian) distribution. Thus, we could simply estimate the mean μ and variance σ^2 in the expression

$$f(x|\mu, \sigma^2) = \frac{1}{\sqrt{2\pi\sigma^2}} e^{-(x-\mu)^2/2\sigma^2}. \tag{4.2.1}$$

Such a belief in a distribution as being "universal" goes back to the 19th century. Francis Galton coined the name "normal" to indicate this universality. He stated (1879): "I know of scarcely nothing so apt to impress the imagination as the wonderful form of cosmic order expressed by the 'Law of Frequency of Error.' The law would have been personified by the Greeks and deified, if they had known of it. It reigns with serenity and in complete self-effacement amidst the wildest confusion. The huger the mob and the greater the apparent anarchy, the more perfect is its sway. It is the supreme law of Unreason."

Galton is here discussing the practical manifestations of the central limit theorem: the fact that if we sum random variables from most practical distributions, then the sum tends to a normal variate. So strong was Galton's belief in normality that in cases where the data were manifestly nonnormal, he assumed that somehow it had been run through a filter before it was observed. It is not uncommon for any of us to decide that a data set which does not conform to our predetermined notions conflicts with the "higher reality" of our own well-cherished biases. Thus, Galton proposed such related distributions as the log–normal; that is, he assumed that the logarithm of the data was normally distributed. Clearly, the transformation to symmetry, which is so important in EDA, is very much in the spirit of Galton.

In most applications, it is very hard to see how the resulting data points are each, in actuality, the result of a summing process that would produce normality. Nevertheless, it is a practical fact that very many data sets either are nearly normal or can be transformed to near normality by a transformation to symmetry. Galton was not naive, even less so was Fisher. Both used the assumption of normality very extensively. Although we can get in serious trouble by assuming that a data set is normal, it seems to be a fact that we get effective normality more often than we have a right to expect.

When data are not normal, what shall we do? One approach might be to seek some sort of transformation to normality or (in practice, almost equivalently) to symmetry. This is very much in the spirit of EDA. If the data can readily be transformed to symmetry, there is still the possibility of contamination by outliers. These may be introduced by the blending in of observations from a second distribution, one that does not relate to the problem at hand, but that can cause serious difficulties if we use them in the estimation of the characterizing parameters of the primary distribution. Or, outliers may be actual observations from the primary distribution, but that distribution may have extremely long tails, for example, the Cauchy distribution. From one point of view, EDA can be viewed as a perturbation approach of normal theory. The data are "massaged" until it makes sense to talk, for example, about a parameter of location (centrality).

Nonparametric density estimation has its primary worth in dealing with situations where the data are not readily transformed to symmetry about a central mode. As such, it is much farther from normal theory than EDA. Although some (e.g., Devroye and Gyorfi, 1985) have developed techniques that are designed to handle outlier problems, the main application of nonparametric density estimation is in dealing with regions of relatively high density. Unlike both classical parametric estimation and EDA, the methodology of nonparametric density estimation is more local and less global.

For example, let us suppose that the data come from a 50–50 mixture of two univariate normal distributions with unit variances and means at -2 and $+2$, respectively. The classical approach for estimating the location parameter would give us a value of roughly 0. The blind use of a trimmed mean approach would also put the location close to 0. But, in fact, it makes no particular sense to record 0 as a measure of 'location." We really need to use a procedure that tells us that there is not one mode, but two. Then, using the two modal values of -2 and $+2$, as base camps, one can gingerly look around these local centers of high activity to get a better glimpse at the structure that generated the data.

Naturally, for low-dimensional data, simply looking at scattergrams would give the user a warning that normal theory (or perturbations thereof) was not appropriate. In such cases, EDA approaches like MacSpin are particularly useful in recognizing what the underlying structure is.

As has been noted in Section 4.1, there are problems in getting the human observer out of the loop for such procedures as MacSpin. Another problem is that in cases where there are a great number of data points, a scattergram does not converge to anything; it simply blackens the page. The scattergram does not exploit continuity in the way that nonparametric density estimation does. It makes sense to talk about consistency with a density estimator. As the data get more and more extensive, the nonparametric density estimator converges to the underlying probability density that characterizes the mechanism that generated the data.

To get to the "nuts and bolts" of nonparametric density estimation, we recall the construction of the histogram. Let us take the range of n univariate

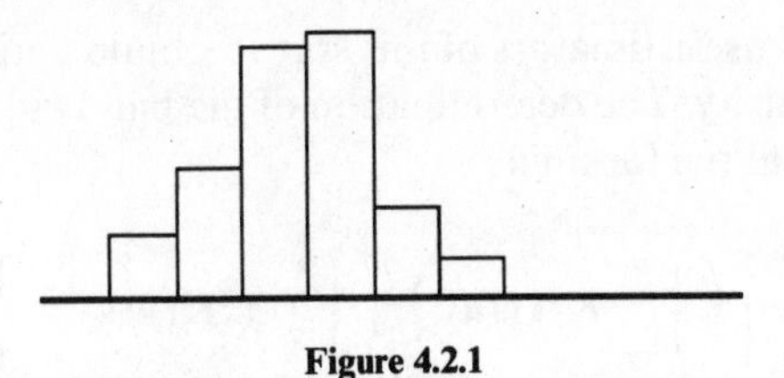

Figure 4.2.1

data points and partition it into m bins of width h. Then the histogram estimate for the density in a bin is given by

$$f_H(x) = \frac{\text{number of data points in bin containing } x}{nh}. \tag{4.2.2}$$

The graph in Figure 4.2.1 shows the kind of shape of the histogram estimator. Clearly, there are disadvantages. The histogram estimator has discontinuities at the bin boundaries, and any naive attempts to use the estimator to obtain derivative information of the underlying density are inappropriate. The mean square error rate of convergence of the estimator is $n^{-2/3}$. A recent paper by Scott (1985) shows how by simply computing 16 histograms, the origin of each shifted from the preceding $h/16$ to the right, and averaging point by point over each of the histograms, many of the undersirable properties of histograms are overcome, while still retaining the rapid computational speed of the histogram estimator.

Next to the histogram (and, significantly, the histogram is still the most used nonparametric density estimator), the most popular nonparametric density estimator is the kernel estimator, proposed first by Rosenblatt (1956) and extended and explicated by Parzen (1962). Here, the estimator at a point x is given by

$$f_K(x) = \sum_{i=1}^{n} \frac{K((x - x_i)/h)}{nh}, \tag{4.2.3}$$

where K is a probability density function and the summation is over the data points $\{x_i\}$. A popular kernel here is Tukey's biweight

$$K(y) = \tfrac{15}{16}(1 - y^2)^2 \quad \text{for } |y| \leq 1. \tag{4.2.4}$$

The order of convergence of the mean square error for most kernels is $n^{-4/5}$. A practical implementation of the kernel estimation procedure (NDKER) is included in the popular IMSL software library.

It is possible to use estimators of this sort to obtain derivative estimates of the underlying density. The determination of the bandwidth h can, in theory, be determined from the formula

$$h = \frac{\left[\left(\int_{-\infty}^{\infty} K^2(y)\,dy\right)\Big/\left(\int_{-\infty}^{\infty} y^2K(y)\,dy\right)^2\right]^{1/5}}{\left(n\int_{-\infty}^{\infty}(f''(y))^2\,dy\right)^{1/5}}. \tag{4.2.5}$$

The problem here is that we do not know f, much less f''. An approach suggested by Scott et al. (1977) is to make a preliminary guess for h, use (4.2.3) to obtain an estimate for f, differentiate it, and plug into (4.2.5). The process is continued until no further change in the estimate for h is observed. A more sophisticated approach for the selection of h has recently been given by Scott and Terrell (1986).

A more local procedure than the kernel estimator is the kth nearest neighbor kernel estimate

$$f(x) = \sum_{i=1}^{n} \frac{K((x - x_i)/d_k(x))}{nd_k(x)}, \tag{4.2.6}$$

where $d_k(x)$ is the distance from x to the kth data point nearest to it. The bandwidth parameter here, of course, is k.

Another estimation procedure is the maximum penalized likelihood approach suggested by Good and Gaskins (1971) and generalized by de-Montricher et al. (1975), Scott et al. (1980), and Silverman (1982). In one of the simple formulations, the procedure finds the f in a large class of functions (say, the class of all probability densities sufficiently smooth that the integral of the square of the second derivative is finite) which maximizes

$$J(f) = \sum_{i=1}^{n} \log[f(x_i)] - \alpha\int_{-\infty}^{\infty}(f''(y))^2\,dy. \tag{4.2.7}$$

Without the penalty term (i.e., when $\alpha = 0$) we can obtain an infinite value of J by simply putting a Dirac function, with weight $1/n$ at each data point. In practice, we can find a useful value of α by simply plotting the density, as in Figure 4.2.2. An implementation (NDMPLE) is given in the IMSL library. The maximum penalized likelihood approach is particularly useful in problems associated with time-dependent processes (e.g., see Bartoszyński et al., 1981). The maximum penalized likelihood approach is not so useful for densities of random variables of dimension greater than two.

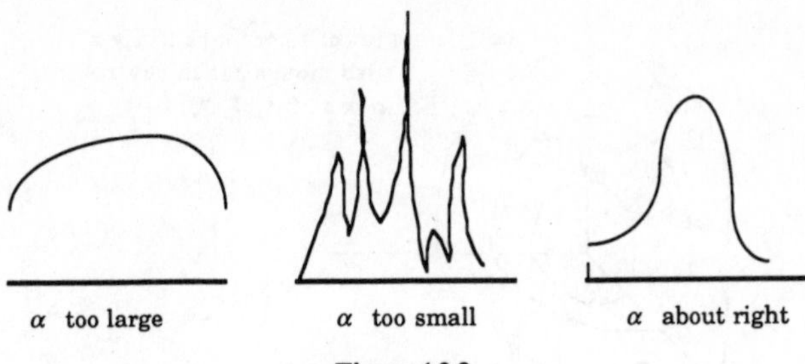

Figure 4.2.2

It is unfortunate that well over 95% of the papers written in the area of nonparametric density estimation deal with the univariate data case, for we now have many procedures to deal with the one-dimensional situation. The problem in the higher-dimensional case is very different from that with one-dimensional data, as we argue below.

Suppose we are given the choice between two packets of information:

A: a random sample of size 100 from an unknown density.

B: exact knowledge of the density on an equispaced mesh of size 100 between the 1 and 99% percentiles.

For one-dimensional data, most of us, most of the time, will opf for option B. However, for four-dimensional data, the mesh in option B would give us only slightly more than three mesh points per dimension. We might find that we had our 100 precise values of the density function evaluated at 100 points where the density was effectively zero. Here we see the high price we pay for an equispaced Cartesian mesh in higher dimensions. If we insist on using it, we shall spend most of our time flailing about in empty space.

On the other hand, the information of packet A remains useful in four-dimensional space, for it gives 100 points that will tend to come from regions where the density is relatively high. Thus, they provide anchor points from which we can examine, in spherical search fashion, the fine structure of the density.

Now we must observe that the criterion of those who deal almost exclusively with one-dimensional data is to transform information of type A into information of type B. Thus, it is very wrong in nonparametric density estimation to believe that we can get from the one-dimensional problem to those of higher dimensionality of a simple wave of the hand. The fact is that "even a rusty nail" works with one-dimensional data. We still know very little about what works for the higher-dimensional problems. Representational problems are dominant. The difficulty is not so much being able to estimate a density function

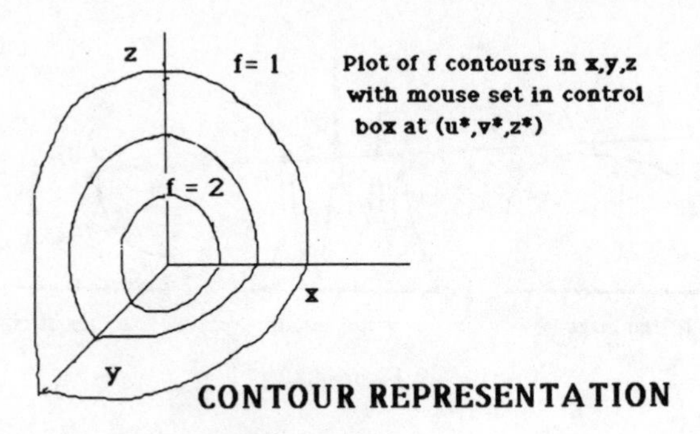

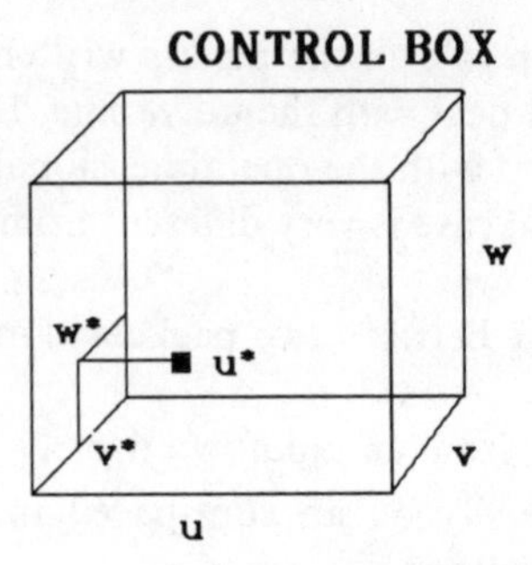

Figure 4.2.3

at a particular point, but knowing where to look. We can, if we are not careful, spend an inordinate amount of time coming up with excellent estimates of zero. We shall discuss two of the more promising avenues of dealing with the higher-dimensional problem next. The first is an attempt to extend what we have learned in density estimation in lower dimensions to higher dimensions, emphasizing graphical display. For example, we see in Figure 4.2.3 (Scott and Thompson, 1983) a display of estimated density contours appropriate for six-dimensional random variables. We shall utilize contours of the boundary of the first three dimensions of our data, which satisfy

$$f(x, y, z) \geq c. \tag{4.2.8}$$

Clearly, as the value of the density function increases, we should expect to see a smaller region which satisfies the condition in (4.2.8). For the next three variables, (u, v, w), we create a control box. As we move the cursor around inside the (u, v, w) box, we note the shape of the density contours inside the (x, y, z) box. This procedure, which is clearly based on utilizing the fundamentally three-dimensional perception of the human eye, is obviously highly interactive. The similarities between such a density estimation approach and

EDA scattergrams are clear. The problem of "fading to black" with large data sets has been eliminated. Moreover, the presence of a "man in the loop" would seem to be less than with the scattergram. The notion of a region having points of density greater than a specified amount can be automated.

A second approach (Boswell, 1983, 1985) is automated from the outset. The objective of the Boswell algorithms is the discovery of foci of high density, which we can use as "base camps" for further investigation. In many situations, the determination of modal points may give us most of the information we seek. For example, if we wish to discriminate between incoming warheads and incoming decoys, it may be possible to establish "signatures" of the two genera on the basis of the centers of the high-density regions.

We shall give a brief glimpse at the simplest of the Boswell algorithms. We are seeking a point of high density, a local maximum of the density function:

(4.2.9) Algorithm 1

$x_c = x_0$

Do until stopping criteria are satisfied

$x_c \leftarrow$ mean of k nearest neighbors of x_c.

In Figure 4.2.4, we sketch the result of (4.2.8) when applied to the estimation of a normal variate centered at zero with identity covariance matrix based on a sample of size 100 for dimensionality (p) through 100. If we look at the standardized (divided by the number of dimensions) mean squared error of the estimate, we note that it diminishes dramatically as p increases to 5 and does not appear to rise thereafter.

Naturally, we need the algorithm to deal with the more complex situation where the number of modes is large and unknown. This has been done with the Boswell approach by making multiple starts of the algorithm (4.2.8), saving

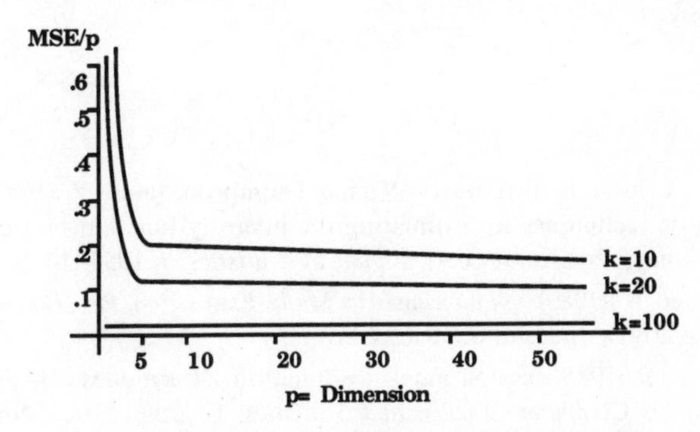

Figure 4.2.4. Estimation of μ from multivariate normals.

the various x_c values in a file, and coalescing the estimated modes into a smaller collection.

(4.2.10) Algorithm 2
For each data point x_i set $x_c = x_i$
Perform Algorithm 1 to produce mode estimate μ_i
Save μ_i in a workfile
End
Analyze the set $\{\mu_i\}$ by repeating Algorithm 2 with the $\{\mu_i\}$ treated as the input data set.

Algorithm 2 appears to perform reasonably well as a technique for finding the modes of mixtures of distributions (e.g., Fisher's iris data).

In summary, the power of the digital computer is now allowing us to consider the kinds of high-dimensional models which were avoided in earlier times for sheer reasons of feasibility. Nonparametric density estimation, together with EDA scattergram analysis, appears to be the major contender for handling higher-dimensional data whose generating density is unknown. Many of the reasonable nonparametric techniques, such as rank tests, are only usable on one-dimensional data. We now have the computing power available to answer some really important questions of multivariate data. For example, what price do we pay for following the usual technique of looking at low-dimensional projections? Ought we to make a serious attempt to deemphasize the Cartesian coordinate system and go to spherical representations for multivariate data? When the data are not unimodal, ought we to move to multiple origin representations rather than single origin representations? How soon can we develop completely automated nonparametric density estimation algorithms for detection purposes? Can we use nonparametric density estimation as an exploratory device to get us back to algorithms based on modified normal theory?

References

Bartoszyński, Robert, Brown, Barry W., and Thompson, James R. (1981). Some nonparametric techniques for estimating the intensity function of a cancer related nonstationary Poisson process, *Annals of Statistics*, **9**, 1050–1060.

Boswell, Steven B. (1983). *Nonparametric Mode Estimation for Higher Dimensional Densities*, Ph.D. dissertation, Rice University.

Boswell, Steven B. (1985). Nonparametric estimation of the modes of high-dimensional densities, in *Computer Science and Statistics*, L. Billard, ed., North Holland, Amsterdam, pp. 217–225.

Devroye, Luc and Gyorfi, L. (1985). *Nonparametric Density Estimation: The L_1 View*, Wiley, New York.

Galton, Francis (1879). *Natural Inheritance*, Macmillan, London.

Good, I. J. and Gaskins, R. A. (1971). Nonparametric roughness penalties for probability densities, *Biometrika*, **58**, 255–277.

Husemann, Joyce Ann (1986). *Histogram Estimators of Bivariate Densities*, Ph.D. dissertation, Rice University.

IMSL (1985). Subroutine NDKER.

IMSL (1985). Subroutine NDMPLE.

deMontricher, Gilbert, Tapia, Richard A., and Thompson, James R. (1975). Nonparametric maximum likelihood estimation of probability densities by penalty function methods, *Annals of Statistics*, **3**, 1329–1348.

Parzen, Emmanuel (1962). On the estimation of a probability density function and mode, *Annals of Mathematical Statistics*, **33**, 1065–1076.

Rosenblatt, Murray (1956). Remarks on some nonparametric estimates of a density function, *Annals of Mathematical Statistics*, **27**, 832–837.

Scott, David W. (1985). Average shifted histograms, *Annals of Statistics*, **13**, 1024–1040.

Scott, David W. and Terrell, George R. (1987). Biased and unbiased cross-validation in density estimation, *Journal of the American Statistical Association*, **82**, 1131–1146.

Scott, David W. and Thompson, James R. (1983). Probability density estimation in higher dimensions, in *Computer Science and Statistics*, J. Gentle, ed., North Holland, Amsterdam, pp. 173–179.

Scott, David W., Tapia, Richard A., and Thompson, James R. (1977). Kernel density estimation revisited, *Nonlinear Analysis*, **1**, 339–372.

Scott, David W., Tapia, Richard A., and Thompson, James R. (1980). Nonparametric probability density estimation by discrete maximum penalized likelihood techniques, *Annals of Statistics*, **8**, 820–832.

Silverman, B. W. (1982). On the estimation of a probability density function by the maximum penalized likelihood method, *Annals of Statistics*, **10**, 795–810.

Silverman, B. W. (1986). *Density Estimation for Statistics and Data Analysis*, Chapman and Hall, New York.

Tapia, Richard A. and Thompson, James R. (1978). *Nonparametric Probability Density Estimation*, Johns Hopkins University Press, Baltimore.

CHAPTER FIVE

Paradoxes and False Trails

5.1. SOME PROBLEMS WITH GROUP CONSENSUS

Almost every country on earth is, at least nominally, a democracy. Even in nondemocratic situations, policy decisions are frequently the result of a group rather than a single individual. A corporation will make policy based on decisions of its board of directors. Even in a totalitarian state, it would be impossible for a single individual to control the government unless he had support from an oligarchy of powerful people. Furthermore, although many decisions are binary (e.g., declare war or do not declare war), most involve a selection from a variety of alternatives. Consequently, it is interesting to examine how the wishes of individuals translate into rational group policy decisions. Intuitively, most of those who have been immersed in a democratic society all their lives expect a kind of smooth transition between individual preferences and policy agreed upon by the groups to which they belong.

Let us suppose that we have a collection of individuals $\{1, 2, \ldots, n\} = G$, where the arabic numerals refer to specific individuals. Let us suppose that the individuals have each ordered their preferences among several possible decisions (e.g., the candidates for a particular office would be an example of such a list of decisions) $D = \{a, b, c, \ldots\}$. We shall indicate by $a >_j b$ that the jth member of the group prefers decision a to decision b. By $a >_G b$, we indicate that the group as a whole prefers a to b; that is, whatever the mechanism used to pool the preferences of the group it arrives at a preference of a over b. Naturally, as we go from individual preferences to group consensus, we do not expect that everyone will get his or her first choice. But we would expect that gross discrepancies between individual preferences and group preferences would be unknown in a democratic society. Even in an oligarchical society, such as the Soviet Union, we would be surprised if, say, the pooling of the preferences of the members of the Politburo favored a policy to which all the members, save one, strongly objected.

The means by which groups of individuals form collective policy are not well understood. We shall consider a variant of Roberts' (1976) formulation of Arrow's demonstration that some apparently very reasonable conditions for group consensus are, in fact, mutually inconsistent. We shall express these conditions as four axioms.

(5.1.1) Suppose that $a >_G > b$ for a particular set of individual preferences. Next, suppose that some of the individual preferences are modified in such a way that the preference for a over any other alternative is either unchanged or is modified in a's favor, and each individual's preferences between a and any other alternative are unchanged. Then the group preference for a over b is maintained.

(5.1.2) Again, suppose $a >_G b$. Next, suppose that some of the individual preferences between alternatives other than a and b are modified, but the preferences between a and b are unchanged. Then, the group preference for a over b is maintained.

(5.1.3) For any pair of alternatives a and b, there is some collection of individual preferences for which $a >_G b$.

(5.1.4) There is no individual who has such influence that if he or she prefers a to b, and every other member of the group ranks b over a, then $a >_G b$.

The axioms appear both reasonable and not very restrictive. For example, (5.1.1) simply states that in a situation where the group favors a over b, if some of the voters increase their enthusiasm for a vis-à-vis some other alternative(s) and none of the voters lessen their enthusiasm for a vis-à-vis any other alternative, then the group maintains its preference for a over b.

Axiom (5.1.3) simply states that the collective opinions of the group have some relevance to the group decision. Thus, for any pair of alternatives a and b, there is some conceivable profile of individual preferences which could lead to a group preference of a over b. If (5.1.3) were not true, then the group would not have the power to establish preferences between all alternatives.

Axiom (5.1.4) states that there is no absolute dictator in the group, no individual whose preference between a and b will dominate the group decision even if all other members of the group have the opposite ordering. We note how very weak this axiom is. The governments of both Nazi Germany and Soviet Russia would satisfy (5.1.4).

Of the three axioms, (5.1.2) is perhaps the most open to question. Let us suppose that there are three candidates for a public office: a Democrat, a Republican, and a Vegetarian. Suppose that there are 101 voters. Fifty of these have the following preference: $R > V > D$. Fifty-one of these have the preference $D > R > V$. By (5.1.2) we should be indifferent to the fact that the antipathy of 50 of the voters is so strong to the Democrat that they prefer the Vegetarian, but the 51 voters who favor the Democrat prefer the Republican to the Vegetarian. Perhaps it would be more reasonable to give the election to the Republican, since the negatives associated with the Democrat are so substantial. The "axiom of the irrelevant alternative" then fails to take into consideration the strength of support that the voters have for one candidate relative to another.

We shall now establish Arrow's impossibility theorem, which states that if G has at least two voters and there are at least three decisions to be considered by the group, then if (5.1.1)–(5.1.3) are satisfied, then (5.1.4) is violated. First, we need to make some definitions. Second, we shall establish two straightforward lemmas and then complete the rest of the proof in more or less conversational mode.

(5.1.5) Let us suppose we have $k \geq 3$ mutually exclusive decisions for which each of $n \geq 2$ voters have ordered preferences. Suppose $\mathscr{P} = (P_1, P_2, \ldots, P_n)$ represents the ordered preferences (profile) of each of the individual voters. Suppose J is a subset of the set of individuals G and a and b are included among the set of decisions. Then if, for all the individuals in J, aP_ib (i.e., $a >_i b$), we say that $\mathscr{P}$ is *J-favored* for (a, b); $\mathscr{P}$ is *strictly J-favored for* (a, b) *if* $b > P_i > a$ for all i not contained in J. J is *decisive* for (a, b) if the fact that all members in J favor a over b implies that the group consensus prefers a to b, that is, $aF(\mathscr{P})b$, where F is the group decision rule (social welfare function). A *minimal decisive set* J is a subset of G which is decisive for some (a, b) and which has the property that no subset of J is decisive for any other pair of decisions.

(5.1.6) **Lemma.** Assume (5.1.1)–(5.1.4). Then J is decisive for (a, b) if and only if there is a set of preferences which is strictly *J-favored* for (a, b) and for which $aF(\mathscr{P})b$.

Proof. If J is decisive for (a, b), then any strictly *J-favored* for (a, b) profile has $aF(\mathscr{P})b$. Next, suppose there is a profile $\mathscr{P}$ that is strictly *J-favored* for (a, b) and for which $aF(\mathscr{P})b$. This means that every voter in subset J prefers a to b. Furthermore, all voters not in J prefer b to a. By (5.1.2) we know that the preferences concerning other decisions are irrelevant. Now let $\mathscr{P}'$ be some other *J-favored* for (a, b) profile. By definition, we know that all the voters in J prefer a to b for this second profile. Thus, the a versus b situation is the same for members of the set J for both profiles $\mathscr{P}$ and $\mathscr{P}'$. Let us now consider the complement of J, $G - J$ for $\mathscr{P}$ and $\mathscr{P}'$. We know that for the profile $\mathscr{P}$, all the voters in $G - J$ prefer b to a. In spite of this fact, by hypothesis, the group prefers a to b (i.e.,

$aF(\mathscr{P})b$). Now in the case of $\mathscr{P}'$ the preferences between a and b can be no worse for a than in $\mathscr{P}$. Thus, by (5.1.1) and (5.1.2) the fact that $aF(\mathscr{P})b$ implies that $aF(\mathscr{P}')b$.

(5.1.7) **Lemma.** Assume (5.1.1)–(5.1.4). Then the entire group is decisive for every (a, b).

Proof. Follows immediately from (5.1.3).

Now let J be a minimal decisive set. Clearly, there must be such a set, since (5.1.7) shows that the entire group G is decisive for every (a, b). We simply remove individuals from G until we no longer have a decisive set. Next, we pick an arbitrary decisionmaker j from the set J. We shall prove that j is an absolute dictator, contrary to (5.1.4).

Suppose that J is decisive for (a, b). Pick another decision c which is neither a nor b. Consider the profile in Figure 5.1.1.

Now, by construction, J is decisive for (a, b). Thus, $aF(\mathscr{P})b$. We note that $\mathscr{P}$ is strictly $\{J - j\}$*-favored* for (c, b). Consequently, if $cF(\mathscr{P})b$, then by (5.1.6) $\{J - j\}$ would be decisive for (c, b), contrary to the assumption that J is a minimal decisive set. Thus, c is not favored over b by the group, and a is favored over b by the group. Consequently, we have two possible scenarios for the preference of the group: a is preferred to b is preferred to c, or a is preferred to b and c, and the group ties b and c. In either case, we conclude that $aF(\mathscr{P})c$. But j is the only individual who prefers a to c. Thus, by (5.1.6), j is decisive for (a, c). Thus, j cannot be a proper subset of J; that is, $j = J$. So far, we have shown that j is decisive for (a, c), for any $c \neq a$.

We now wish to establish that the individual decisionmaker j is decisive for (d, c) for any d, c not equal to a. Consider the profile in Figure 5.1.2.

P_i for $i \in J - \{j\}$	P_i for $i \notin J$	P_j
c	b	a
a	c	b
b	a	c
$D - \{a, b, c\}$	$D - \{a, b, c\}$	$D - \{a, b, c\}$

Figure 5.1.1

P_j	P_i for $i \neq j$
d	c
a	d
c	a
$D - \{a, c, d\}$	$D - \{a, c, d\}$

Figure 5.1.2

P_j	P_i for $i \neq j$
d	c
c	a
a	d
$D - \{a, c, d\}$	$D - \{a, c, d\}$

Figure 5.1.3

Now the entire group G is decisive for any pair of decisions. Hence, $dF(\mathscr{P})a$. Now, since j is decisive for (a, c), we have $aF(\mathscr{P})c$. Thus, the group ranks d over a and ranks a over c. So $dF(\mathscr{P})c$. Consequently, by (5.1.6), j is decisive for (d, c).

Finally, we shall demonstrate that j is decisive for (d, a) whenever $d \neq a$. Consider the profile in Figure 5.1.3.

Since j is decisive for (d, c), we have $dF(\mathscr{P})c$. Since the entire group is decisive (by (1.5.7)), we have $cF(\mathscr{P})a$. Consequently, $dF(\mathscr{P})a$. But j is the only individual preferring d to a. Thus, we have established that j is an absolute dictator, contrary to (1.5.4). In summary, we have established the following theorem.

Arrow's Impossibility Theorem. If (1.5.1)–(1.5.3) hold, then (1.5.4) cannot hold if there are at least two decisionmakers and more than two possible decisions.

What, indeed, should be our response to this rather surprising result? Some hold that it raises real questions about democracy in particular and group decisionmaking in general. Yet the empirical fact is that groups do appear to make decisions in a more reasonable fashion than Arrow's theorem would lead us to believe. Does this contradiction between a rational argument and empirical observation imply some kind of necessity for the suspension of reason or a denial of our common sense? It should cause us first of all to examine Arrow's model very carefully to see how reasonable the axioms are. It is too glib to point out that in the Anglo-Saxon democracies there are generally only two major parties, so that Arrow's theorem does not apply. There are, generally speaking, primaries in the United States in which more than two candidates are considered. In Great Britain, most seats are contested by candidates from three parties. Furthermore, democracy appears to work well in the Scandinavian countries and in Israel, for example, where offices in general elections are generally contested by more than two candidates. Moreover, commercial boards of directors usually are confronted with decisions with more than two alternatives. The answer(s) lies elsewhere.

As we have seen, an acceptance of "the axiom of irrelevant alternatives (5.1.2)" implies that the strength of preferences by the individual voters is not taken into consideration. This is clearly erroneous in most situations. In the American presidential election of 1968, many in the left wing of the Democratic party refused to support Senator Humphrey, because of his involvement with the policy of President Johnson in the Vietnam war. This dissent was essential in the election of President Nixon. It was not that the dissidents preferred Richard Nixon to Hubert Humphrey or that they assumed that their vote for ad hoc third party candidates would be successful. They wished to make a statement which, they rightly assumed, would have consequences in the molding of policy of the Democratic party in the future. Thus, we observe that the notion of a single decision independent from other decisions is generally flawed. People approach the totality of their decisionmaking tasks in a complex, highly interrelated fashion, which simply has not yet been satisfactorily modeled. In the case of the election of 1968, the strategy of the left wing of the Democratic party begun then influences elections to this day. By this strategy, the left wing of the party essentially forces aspirant candidates to adopt positions that are well to the left of those of the general electorate. This has generally helped Republicans in presidential elections, but, from the standpoint of the Democratic extreme left, there is essential indifference between the policies of center-right Democrats and those of the Republican party. Thus, from the standpoint of the Democratic left, their strategy, begun in 1968, gives them an influence in their party far out of proportion to their numbers. Such strategies have no place in Arrow's system, but they are quite reasonable and reflect the real world much better than do Arrow's axioms.

Whenever we axiomitize a real-world system, we always, of necessity, oversimplify. Frequently, the oversimplification will adequately describe the system for the purposes at hand. In many other cases, the oversimplification may seem deceptively close to reality, when in fact it is far wide of the mark. The best hope, of course, is the use of a model adequate to explain observation. However, when we are unable to develop an adequate model, we would generally be well advised to stick with empiricism and axiomatic imprecision. The fact is that Arrow's theorem is more useful as a shock to stress our notions of group decisionmaking than it is an excuse to change the structure of democracy or the means of creation of boardroom policy.

In a number of state legislatures and in the United States House of Representatives, electronic voting systems exist which enable instantaneous computerized voting. The original idea was not only to improve the accuracy of enumeration of votes, but also to eliminate the boring and time-consuming process of vote roll calling. This latter feature, the instantaneous voting by all members, has caused problems. Why? The traditional sequencing of voting, where legislators have the right to defer the time in the order in which they vote enables a kind of passive coalition building where the legislators who feel most strongly about a bill speak up first. It is not that the voting device fails to do what was promised. It is rather that the axioms governing the construction of the device do not conform to reality.

As a practical matter, the formal rules of voting probably are not terribly important. Many nations have written constitutions quite as impressive in the reading as that of the United States. Most of these are de facto dictatorships or oligarchies. The United Kingdom has no written constitution, and many of the matters formalized in the United States are dealt with by "gentlemen's agreement" in Britain. Yet the functioning of both nations is very similar, and although collectivization has proceeded further in Britain than in the United States, few would argue that the differences in democratic forms in the two nations have had a major impact on the pervasiveness of the state in matters of welfare and private ownership. The notion of "a government of laws, not of men" sounds good, but history shows that it is temperament and other unwritten and difficult to define concepts which are much more important than formal rules of voting and governance. Nevertheless, the forms are not without some importance, in the sense that some forms of voting may be appropriate for one decisionmaking group and not for another.

At the level of the board of directors of a firm, decisions are generally by unanimous agreement. The structure of an effective board is oriented to a "cluby," nonadversarial means of consensus. The board members are elected "at large" by the stockholders so that it is unlikely that members holding diametrically opposing views will be selected. This is particularly the case, since members of the board generally make recommendations to the stockholders about who should be appointed to a newly opened position on the board. Frequently, decisions at the boardroom level are really made by one individual who has special expertise and/or concern about the matter at hand. For example, suppose that a decision is to be made concerning the use of a new microchip for the central processing unit of a computer. Generally, only one or two individuals on the board will have expertise in the area, and generally the subset of individuals with expertise will have agreed on a common decision prior to the formal sitting of the board. The other members of the board are unlikely to vote against their expert members, since, presumably, a major reason for their selection to the board was their microchip expertise. Suppose there are 12 members on the board. Prior to the meeting, all the 11 nonexpert members had positive feelings about the use of a particular chip, since they had read a favorable article about it in *The Wall Street Journal.* These feelings are not strongly held, since they are based on little information. If the member of the board who does have microchip expertise comes out strongly against the chip, his colleagues will almost certainly support him. All the members of the board have as their constituency the stockholders. The stockholders have a common goal: to make money. The task of the board is to implement whatever policies will achieve this result. If there should be one or more members of the board who find themselves in constant opposition to the policy of the majority, they will generally either be co-opted by the majority, resign from the board, or (very rarely) attempt to have the stockholders replace the majority members with others who agree with the agenda of the current minority. As a very practical matter, the voting structure of boardrooms is so dominated by nonantagonism and clubiness that boards are in constant danger of becoming unimaginative and stifling of innovation unless there is a truly competitive market situation that keeps the board attentive to its task.

Civic decisionmaking is quite different. There is no well-defined bottom line for a city, state, or nation. There are goals on which almost all would agree, such as defense from foreign invasion. But other goals, such as government health insurance, will not be shared by all. Consequently, in effective democracies, there will be a constant necessity to see to it that decisions are made which will be accepted even by those who may strongly oppose them. A major way of achieving this goal is to see to it that no law is passed which will be deemed "over the top of the wall" by minorities or even by individuals. In the 17th century, the Scottish parliament, enraged at an alleged act of murder committed by members of the Clan MacGregor, actually passed a law making it a capital offense for any MacGregor to exist. This law was finally rescinded when members of other clans understood that such a law could be written to apply to them too.

A second means, much less important than the first, of achieving the goal of compliance is the construction of formal voting mechanisms which will be deemed "fair." If, in the Parliamentary election of 1987, all members had stood "at large," then there can be little doubt that the current House of Commons would be nearly 100% Conservative, since so overwhelming a majority of individuals in the south of England voted Conservative that it translated into a large national majority for the Conservatives. No reasonable member of the Conservative party would advocate such a means of voting, even though it would guarantee easy passage of the programs advocated by Mrs. Thatcher. Majorities in Scotland and the north of England strongly favored Labour. Were the people in these areas of Britain denied representation in the Commons, separatist movements would blossom overnight. Thus, members of Parliament do not run at large but stand for election from relatively small geographical areas. Practically every country has some similar device for "bringing government closer to the people." The decisions that will be made by the current House will be very similar to those that would have been made if the members had run for office at large. Yet, due to the perceived "fairness" of the voting system, compliance with the decisions can be expected to be much greater than would have been the case otherwise.

In the United States, the existence of the states creates a problem of allocation of seats in the House of Representatives which has no parallel in the United Kingdom. Namely, congressional districts cannot cross state boundaries. The number of congressional seats allocated to each state is such that each state is guaranteed at least one seat; otherwise, the number for each state "shall be apportioned... according to their respective Numbers...." An extensive mathematical analysis of congressional apportionment has been given by Balinski and Young (1982). Almost every American "knows" how the number of seats assigned to each state is determined. The method they believe operational is that of Alexander Hamilton, which is the following:

(5.1.8) Pick the size of the House $= n$. Divide the voting population N_j of the jth state by the total population N to give a ratio r_j. Multiply this ratio by n to give the quota q_j of seats for the jth state. If this quota is less than one, give the state one seat. Give each state the number of representatives equal to the integer part of its quota. Then rank the remainders of the quotas in descending order. Proceeding down the list, give one additional seat to each state until the size of the House n has been equaled.

The method of Hamilton has firmly embodied in it the notion of the state as the basic entity of indirect democracy. Once the number of representatives has been arrived at by a comparison of the populations of the several states, the congressional districts could be apportioned by the state legislatures within the states. But the indivisible unit of comparison was that of state population. If we let a_j be the ultimate allocation of seats to each state, then it is obvious that for all $p \geq 1$, the method of Hamilton minimizes

$$\sum_{j=1}^{k} |a_j - q_j|^p, \tag{5.1.9}$$

where k is the number of states.

Let us consider the results of employing the method of Hamilton based on the population figures utilized in the first allocation year of 1792 in which it was proposed to fix the size of the House that year at 120 (see Table 5.1.1).

It is interesting to note that the first congressional Senate and House passed a bill embodying the method of Hamilton and the suggested size of 120 seats. That the bill was vetoed by President George Washington and a subsequent method, that of Thomas Jefferson, was ultimately passed and signed into law is an interesting exercise in Realpolitik. The most advantaged state by the Hamiltonian rule was Delaware, which received a seat for every 27,770 of its citizens. The most disadvantaged was Georgia, which received a seat for every 35,417 of its citizens. Jefferson's Virginia was treated about the same

Table 5.1.1. The Method of Hamilton

State	Population	Quota	Hamiltonian Allocation	Voters per Seat
Connecticut	236,841	7.860	8	29,605
Delaware	55,540	1.843	2	27,770
Georgia	70,835	2.351	2	35,417
Kentucky	68,705	2.280	2	34,353
Maryland	278,514	9.243	9	30,946
Massachusetts	475,327	15.774	16	29,708
New Hampshire	141,822	4.707	5	28,364
New Jersey	179,570	5.959	6	29,928
New York	331,589	11.004	11	30,144
North Carolina	353,523	11.732	12	29,460
Pennsylvania	432,879	14.366	14	30,919
Rhode Island	68,446	2.271	2	34,223
South Carolina	206,236	6.844	7	29,462
Vermont	85,533	2.839	3	28,511
Virginia	630,560	20.926	21	30,027
Total	3,615,920	120	120	

as Hamilton's New York with a representative for every 30,027 and 30,144 citizens, respectively. The discrepancy between the most favored state and the least favored was around 28%, a large number of which the supporters of the bill were well aware. The allocation proposed by Hamilton in 1792 did not particularly favor small states. In general, however, if we assume that a state's likelihood of being rounded up or down is independent of its size, the method of Hamilton will favor somewhat the smaller states if our consideration is the number of voters per seat. But Hamilton, who was from one of the larger states and who is generally regarded as favoring a strong centralized government that deemphasized the power of the states, is here seen as the clear principled champion of states' rights and was apparently willing to give some advantage to the smaller states as being consistent with, and an extension of, the notion that each state was to have at least one representative.

Now let us consider the position of Thomas Jefferson, the legendary defender of states' rights. Jefferson was, of course, from the largest of the colonies, Virginia. He was loathe to see a system instituted until it had been properly manipulated to enhance, to the maximum degree possible, the influence of Virginia. Unfortunately for Jefferson, one of the best scientific and mathematical minds in the United States who undoubtedly recognized at least an imprecisely stated version of (5.1.9), the result in (5.1.9) guaranteed that there was no other way than Hamilton's to come up with a reasonable allocation rule fully consistent with the notion of states' rights. Given a choice between states' rights and an enhancement of the power of Virginia, Jefferson came up with a rule that would help Virginia, even at some cost to his own principles.

Jefferson arrived at his method of allocation by departing from the state as the indivisible political unit. We consider the method of Jefferson below.

(5.1.10) Pick the size of the House = n. Find a divisor d so that the integer parts of the quotients of the states when divided by d sum to n. Then assign to each state the integer part of N_j/d.

We note that the notion of a divisor d is an entity that points toward House allocation which could occur if state boundaries did not stand in the way of a national assembly without the hindrance of state boundaries. Let us note the effect of Jefferson's method (Table 5.1.2) using the same census figures as in Table 5.1.1.

We note that the discrepancy in the number of voters per representative varies more with Jefferson's method than with Hamilton's—94% versus 28%. In the first exercise of the presidential veto, Washington, persuaded by Jefferson, killed the bill embodying the method of Hamilton, paving the way for the use of Jefferson's method using a divisor of 33,000 and a total House size

Table 5.1.2. The Method of Jefferson (Divisor of 27,500)

State	Population	Quotient	Jeffersonian Allocation	Voters per Seat
Connecticut	236,841	8.310	8	29,605
Delaware	55,540	1.949	1	55,540
Georgia	70,835	2.485	2	35,417
Kentucky	68,705	2.411	2	34,353
Maryland	278,514	9.772	9	30,946
Massachusetts	475,327	16.678	16	29,708
New Hampshire	141,822	4.976	4	35,456
New Jersey	179,570	6.301	6	29,928
New York	331,589	11.635	11	30,144
North Carolina	353,523	12.404	12	29,460
Pennsylvania	432,879	15.189	15	28,859
Rhode Island	68,446	2.402	2	34,223
South Carolina	206,236	7.236	7	29,462
Vermont	85,533	3.001	3	28,511
Virginia	630,560	22.125	22	28,662
Total	3,615,920	120	120	

Table 5.1.3. Allocations of the Methods of Hamilton and Jefferson

State	Population	H	J	Voters/Seat H	Voters/Seat J
Connecticut	236,841	7	7	33,834	33,834
Delaware	55,540	2	1	27,220	55,440
Georgia	70,835	2	2	35,417	35,417
Kentucky	68,705	2	2	34,353	34,353
Maryland	278,514	8	8	34,814	34,814
Massachusetts	475,327	14	14	33,952	33,952
New Hampshire	141,822	4	4	35,455	35,455
New Jersey	179,570	5	5	35,914	35,914
New York	331,589	10	10	33,159	33,159
North Carolina	353,523	10	10	35,352	35,352
Pennsylvania	432,879	13	13	33,298	33,298
Rhode Island	68,446	2	2	34,223	34,223
South Carolina	206,236	7	7	29,462	29,462
Vermont	85,533	2	2	42,766	42,776
Virginia	630,560	18	19	35,031	33,187
Total	3,615,920	105	105		

of 105. Let us examine the differences between the method of Hamilton and that of Jefferson (see Table 5.1.3).

The only practical difference between the two allocation systems is to take away one of Delaware's two seats and give it to Virginia. The difference between the maximum and minimum number of voters per seat is not diminished using the Jeffersonian method, which turns out to give a relative inequality of 88%; for the Hamiltonian method the difference is a more modest 57%. The method of Jefferson favors the larger states pure and simple. But the difference was not enough to persuade the Congress to spend more time on what is, in reality, a minor advantage. They approved the method of Jefferson, and this method was in use until after the census of 1850, at which time the method of Hamilton was installed and kept in use until it was modified by a Democratic Congress in 1941 in favor of yet another scheme.

We have in the 1792 controversy a clear example of the influence on consensus of one individual who passionately and cleverly advances a policy about which his colleagues have little concern and less understanding. Jefferson was the best mathematician involved in the congressional discussions, and he sold his colleagues on a plan in the fairness of which one may doubt he truly believed.

Naturally, as the large state bias of the method of Jefferson began to be understood, it was inevitable that someone would suggest a plan that, somewhat symmetrically to Jefferson's method, would favor the small states. We have such a scheme proposed by John Quincy Adams.

(5.1.11) Pick the size of the House $= n$. Find a divisor d so that the integer parts of the quotients (when divided by d) plus 1 for each of the states sum to n. Then assign to each state the integer part of $N_j/d + 1$.

The plan of Adams gives the same kind of advantage to the small states that that of Jefferson gives to the large states. Needless to say, it has never been used in this country or any other (though, amazingly, Jefferson's has).

It is interesting to note that Daniel Webster attempted to come up with a plan that was intermediate to that of Jefferson's and that of Adams. He noted that whereas Jefferson rounded the quotient down to the next smallest integer, Adams rounded up to the next largest integer. Webster, who was a man of incredible intuition, suggested that fractions above 0.5 be rounded upward, those below 0.5 be rounded downward. From the 1830s until 1850 there was very active discussion about the unfairness of the method of Jefferson and a search for alternatives. It was finally decided to pick Hamilton's method, but Webster's was almost selected and it was a contender as recently as 1941. As it turns out, there is very little practical difference between the method of Hamilton and that of Webster. Both methods would have given identical allocations from the beginning of the Republic until 1900. Since that time, the differences between the two methods usually involve one seat per census.

The method of Hamilton was replaced in 1941 by one advocated by Edward Huntington, Professor of Mathematics at Harvard. Instead of having the division point of fractions rounded up and rounded down to one-half, Huntington advocated that if the size of the quotient of a state were denoted by N_j/d then the dividing point below which rounding down would be indicated would be the geometric mean $\sqrt{[N_j/d]([N_j/d] + 1)}$, where $[\cdot]$ denotes "integer part of." One might say that such a method violates the notion that such methods should be kept simple. Furthermore, the rounding boundaries do increase slightly as the size of the state increases, giving an apparent advantage to the smaller states. At the last minute, the more popular method of Webster was rejected in favor of that of Huntington, since its application using the 1940 census would give a seat to Democratic Arkansas rather than to (then) Republican Michigan. The Huntington method is in use to this day, though not one American in a thousand is aware of the fact.

And indeed, it is not a very important issue whether we use the method of Hamilton or that of Webster or that of Huntington or even that of Jefferson or that of Adams. Not one significant piece of legislation would have changed during the course of the Republic if any one of them were chosen. The subject of apportionment possibly receives more attention than practicality warrants.

Another matter, which is incredibly more important, is that of gerrymandering. Once a state has received an allocation of representatives following the official census, which occurs on years divisible by 10, the state legislature then picks the boundary lines of the congressional districts. Federal law is interpreted as requiring nearly equal within-state district size. However, the legislature is free to pick boundaries that are convenient to the party that controls the state house. Suppose that a state is 50% Republican and 50% Democratic. Suppose further that the state has been awarded a total of 10 seats, each of size 500,000. One might suppose that an equitable allocation would be one that gave the Republicans five seats and the Democrats five. On the other hand, if the Democrats happen to be in control of the state house, they might divide the district boundaries so that one of them has 400,000 Republicans and the other nine have 233,333 each. Thus, the Republicans would wind up with one seat, the Democrats with nine. The situation is seldom so extreme, but if gerrymandering were somehow abolished, the party change in the House of Representatives could, at times, be in excess of 40 seats. In the election of 1980, for example, the total of votes cast for Republicans standing for the House of Representatives was greater than that for the Democratic candidates. Yet the result of the election was a significant Democratic majority.

Unlike the difficulties associated with congressional apportionment, it is easy to point out a number of cases in the history of the Republic where gerrymandering has been highly significant in important legislation. For example, Ronald Reagan favored drastic relaxation of federal involvement in many social programs. He started his administration with control of the presidency and control of the Senate, but with a gerrymandered Democratic majority in the House. Since it was impossible to get the kinds of drastic social spending cuts he favored through a Democratic House, he had to content himself with a Fabian policy. Consequently, he proposed massive reductions in the rate of increase of federal taxation, to which many of the Democratic representatives had to consent as a matter of practical politics. He trusted that this would force the House to agree to social spending cuts in order to avoid massive federal deficits. To some extent, he was successful. However, the Democrats were ultimately able to convince much of the electorate that it was the President rather than they who were responsible for the deficits due to his drastically increased defense budget (which passed the House, again as a result of its popularity with the electorate). Revolutionary changes in the American federal government are possible only when, as did President Roosevelt, the Executive's party controls both the Senate and the House. Consequently, the gerrymandering was instrumental in preventing effective radical change in national welfare policy. This is only one example where gerrymandering has changed the course of history. There are many others. Perhaps one could argue that gerrynmandering is a positive force, since it provides a kind of governmental inertia. But if it is desired to implement the fairly current wishes of the electorate, certainly gerrymandering is a mighty detriment to changing the status quo.

The simple rule of having all candidates run at large could, if the parties intelligently mustered their forces on election day, lead to an even more lopsided result than gerrymandering. That the gerrymandering problem receives relatively less attention from modelers than does the apportionment problem is due to the fact that gerrymandering is much more difficult, much more evasive of quantitative axiomitization.

On the other hand, there are factors that should simplify the elimination of the gerrymandering problem. One is that a truly random allocation should be acceptable to all. We do not actually have to optimize anything, as we essentially do in the allocation problem. If a random allocation, perceived as random by all, actually arrived at a districting which led, in our earlier example, to nine Democratic and one Republican districts, it would probably be acceptable although it is unlikely that a truly random scheme would lead to any such configuration.

Naturally, a random scheme could be achieved by simply listing the voters and using a random number generator (with seeds picked by a mechanism that did not allow for diddling). This would achieve the objective of random assignment. Unfortunately, just as states have their own particular concerns which are a function of place, just so there may be concerns within a state which are location and community based. The concems of citizens in the Texas Panhandle will probably not be identical to those of citizens in Galveston. Consequently, any allocation that does not require that districts be geographically based and connected is unlikely to be acceptable. (It should be noted, however, that the present gerrymandering schemes in Texas sometimes achieve connectivity by joining one mass of Republicans with another hundreds of miles away via a narrow corridor.) It is this geographical connectivity constraint which renders the easy randomization fix to gerrymandering difficult.

Let us consider whether we might not have some reasonable suggestions for getting out of the gerrymandering trap. First, we understand that we may not be able to "start from scratch." The new districts from census to census should bear some relationship to each other. Let us suppose that each voter is characterized by the geographical location of his residence. Thus, each voter is characterized by (x_i, y_i), his or her location of place. It is an easy matter to determine the weighted center of location of the old districts given the new census via

$$(\bar{x}, \bar{y}) = \left(\frac{1}{N} \sum_{i=1}^{N} x_i, \frac{1}{N} \sum_{i=1}^{N} y_i \right), \tag{5.1.12}$$

where N is the population size of a district.

Suppose we wish to design a new set of districts of the same numerosity (k) as the old ones and using as our design centers the centers of the old districts. Let us suppose that we start our design from the k centers

$$(\bar{x}_i, \bar{y}_i)_{i=1,2,\ldots,k}. \tag{5.1.13}$$

We then proceed as follows:

(5.1.14) Pick a random order of the integers $\{1, 2, \ldots, k\}$. Starting from the first of the order, select the member of the total population, not already selected, who is closest to the old mean of that group. In the event of ties, select from the tied members randomly. Continue in this fashion, selecting a new ordering at each round, until the entire population of the state has been exhausted.

The algorithm mentioned above will eliminate the gerrymandering problem and give a fairly selected group. However, it has a problem. Namely, a certain proportion of the voters will be found in fuzzy transition areas between districts. We indicate such an allocation in Figure 5.1.4.

We note that whereas most voters will be in single-hatched areas, the voters in the cross-hatched area will sometimes be from one district and sometimes

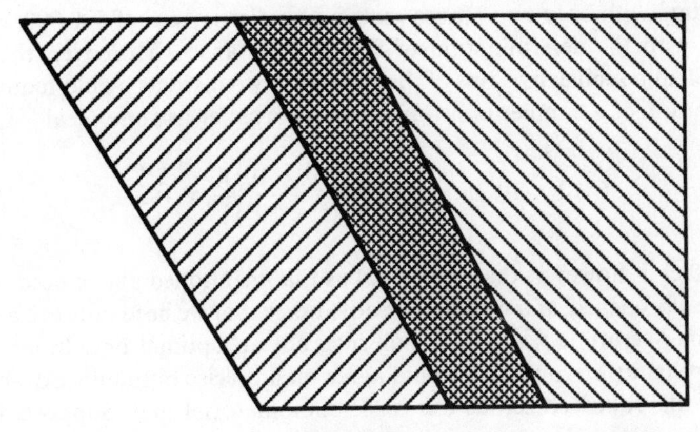

Figure 5.1.4

from another. Indeed, there is no doubt that some transitional areas will exist where voters in a transitional area may be from more than two districts. As a practical matter, such an apparent ambiguity may be of little consequence. It is true that voters in the multiply hatched areas will have a particular bonus in that their interests will be protected in Washington by more than one representative. On the other hand, a representative may well pay less attention to an area where he or she is only partially responsible for the electorate. We recall that each voter will still be allowed to vote for only one congressional seat. Compared to the enormities currently occurring from gerrymandering, the apportionment rule above appears relatively benign.

In order to eliminate the transitional areas above, a number of other computational devices might be employed. For the voters in a transitional area between two districts, we might, for example, agree to exchange two voters from their two districts, if the resulting exchange gave a smaller total distance of the voters from the centers of their districts. This rule would not be unambiguous and would require further randomization, but it could be made reasonable and would reduce the numbers of individuals in the transition area. Also, there is no difficulty in making the kinds of modification required to handle an increase or decrease in the number of districts within a state.

Other, more computationally difficult, rules could be proposed. For example, let us compute, for each voter in the state, the geographical distance of the voter from each other voter. Then, if the sizes of congressional districts in the state are to be of size N, we compute the arrangement which minimizes the average of the maximum distances that voters are from any other voter in the district and do so in such a way that no citizen from a district is surrounded by citizens from any other district. To carry out such a computation naively would require, first of all, the determination of $(kN)^2/2$ distances. For a state with only two House seats and a district size of 500,000, we are dealing with the determination of 5×10^{11} distances. Moreover, the total listing of all possible divisions of the state into two districts would require the enumeration of a number of listings given by a super-astronomical

$$C(10 \times 10^5, 5 \times 10^5) = \frac{(10 \times 10^5)!}{((5 \times 10^5)!)^2} \approx \frac{2^{10^6}}{\sqrt{5\pi}}. \tag{5.1.15}$$

The fact is, however, that such rules as that mentioned above need not be naively approached, and we have the advantage that we need not seek a sharp optimum. We are looking for a fair rule, not an optimal one. Indeed, it is unlikely that any two people would agree on a precise optimum. So we need not use the polled citizen as our basic combinatorial unit. Suppose, in the above, that we decide to use existing voting districts as our basic unit. Suppose that districts have on the average around 5000 citizens within their boundaries. We list all the voting districts in the state and then pick one at random. Proceeding randomly to adjacent districts, we add to the selected district an adjacent district. We then keep adding districts in this fashion to the already formed "superdistrict" until we reach the first number in excess of 50,000 total citizens or are blocked from further amalgamation by being completely surrounded by other superdistricts. Then we select another "seed" district at random and build up another "superdistrict." We continue in this fashion until every district is part of an amalgamated superdistrict (some if these, of course, will have a size equal to just one district, since some of the later random seeds will be surrounded by superdistricts already formed).

We note in the above two-district example that if we have roughly 20 superdistricts, the number of distance computations from the polling centers is simply $20^2/2 = 200$. Once the superdistricts have been formed, the problem of finding the distance measures of all possible combinations is no longer that in (5.1.15) but a manageable

$$C(20,10) = \frac{20!}{(10!)^2} \sim \frac{e^{-20}20^{20}\sqrt{2\pi 20}}{(e^{-10}10^{10}\sqrt{2\pi 10})^2} \approx 1.87 \times 10^5. \tag{5.1.16}$$

There are other problems, naturally, since we will need to form districts that come within some tolerance of 500,000 (for the above example). All these problems can be addressed and solved. There probably is no excuse for gerrymandering in this epoch of sophisticated computing and combinatorics unless it be that, for some political reason, it is desired to maintain it.

The various means of evaluating a vote in particular elections can be quite diverse. A common rule in the Western democracies is that it is desirable to make the rule simple enough that the electorate in general will not feel that it is being manipulated. That rule most familiar in the West is that of the *simple majority*. By this rule, a candidate wins the election if he receives over 50% of the vote. In cases where there are more than two candidates, it may be possible that no candidate receives a majority. In such a case, it is common to have a *run-off* between the top two contenders. Whoever gets the most votes in a run-off wins the election (even if, for some reason, it is not a majority of the votes). In elections where it is impractical to have a run-off, a candidate may be awarded the election by the *plurality rule;* that is, he wins if he receives more votes than any other candidate. Simple rules may be accepted by an electorate even if they are multistaged. For example, in the United States, voters vote for presidential electors, who then elect the president using a majority rule. This is accepted by most as being uncontrived and is, consequently, accepted. If, however, no candidate receives a majority of electoral votes, then the election is thrown into the House of Representatives *where*

P_1	P_2	P_3	P_4
a	b	c	e
b	d	d	d
c	c	a	a
d	e	e	b
e	a	b	c

Figure 5.1.5

P_1	P_2	P_3	P_4
b	d	d	d
c	c	a	a
d	e	e	b
e	a	b	c

Figure 5.1.6

each state would have only one vote. Due to the wide disparity of sizes of populations of the several states, such an eventuality could lead to a crisis of confidence in the equity of the system.

Naturally, there are other, more complicated, systems of voting which are sometimes used in multicandidate situations. For example, let us consider the situation when four voters have the preferences listed in Figure 5.1.5.

According to the *Borda count* rule, each candidate receives a count for each candidate ranked below him. Thus, $B(a) = 4 + 0 + 2 + 2 = 8$; $B(b) = 3 + 4 + 0 + 1 = 8$; $B(c) = 2 + 2 + 4 + 0 = 8$; $B(d) = 1 + 3 + 3 + 3 = 10$; and $B(e) = 0 + 1 + 1 + 4 = 6$. The candidate with the largest Borda count wins. In the above, candidate d, who is no one's first choice, wins the election.

Another rule is "the last shall be first" rule. According to this rule, we take the candidate who receives the smallest number of first choices and treat his second choice as a first choice. In the above, there is a four-way tie for first place, so we go immediately to the second choices and continue the process until a winner results (see Figure 5.1.6).

Again, candidate d wins the election. Such rules have a kind of consensus building appeal but are seldom used due to their complex, nonintuitive structure.

References

Arrow, Kenneth J. (1950). A difficulty in the concept of social welfare, *Journal of Political Economy*, **58**, 328–346.

Balinski, Michel L. and Young, H. Peyton (1982). *Fair Representation: Meeting the Ideal of One Man, One Vote,* Yale University Press, New Haven, CT.

Roberts, Fred S. (1976). *Discrete Mathematical Models,* Prentice-Hall, Englewood Cliffs.

5.2. STEIN'S PARADOX

Let us consider the problem where we need to estimate the mean μ_1 of one normal distribution on the basis of one observation, x_1. The intuitive estimator is simply x_1 itself. We know (see Appendix A.1) that this estimator has expectation μ_1 and variance σ_1^2, where these are the mean and variance of the distribution. Indeed, in this one-dimensional case, it can be shown (Lehman, 1949) that no other estimator has uniformly smaller mean square error ($E(x_1 - \mu_1)^2$) than x_1. If we wished to estimate the mean μ_2 of another normal distribution independent of the first on the basis of one observation, x_2, then again the intuitive estimator would be x_2. If we wished to estimate several means $(\mu_1, \mu_2, \ldots, \mu_k)$ of several independent normal distributions on the basis of one observation from each distribution, $(x_1, x_2, \ldots, x_k)$, then the natural estimator would appear to be $(x_1, x_2, \ldots, x_k)$. If someone would tell us that by shrinking the natural estimator to $(0, 0, \ldots, 0)$, we would be able to do uniformly better than if we used the natural estimator, then we would have strong reason for skepticism. The implications of such a result become even more incredible when we consider that if it were so, it would imply that we could, by translation, pick any other point in k-space to be the magic origin toward which we shrink. If such a result could be proved, then it would be paradoxical indeed. In 1961, James and Stein proved such a result.

Without going into much mathematical detail, we shall indicate why the following version of the Stein result holds. Suppose we wish to estimate the mean of a k-dimensional normal distribution with covariance matrix $\sigma^2 I$ on the basis of an observation $X = (x_1, x_2, \ldots, x_k)$. Then, if we use the loss function $L(\mu^*, \mu) = \sum_i (\mu_i^* - \mu_i)^2/k$, the usual estimator X has uniformly larger risk (the expectation of the loss function) than some estimators of the form $\mu^{**} = g(X^T X)X$, where g is an appropriately chosen function nondecreasing between 0 and 1; that is,

$$Q(\mu^{**}) = E[L(\mu^{**}, \mu)] \leq \sigma^2. \tag{5.2.1}$$

We sketch the result in Figure 5.2.1. There are no mistakes in the James–Stein proof. Given the assumptions made (and normality is not the key driver of the result), the James–Stein theorem is correct.

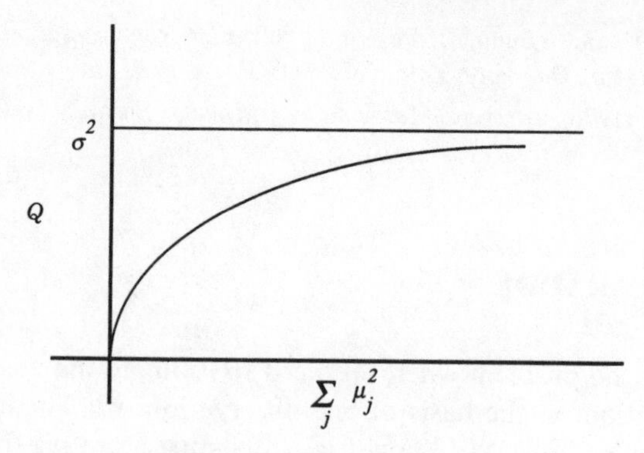

Figure 5.2.1. Risk versus $\sum_j \mu_j^2$.

What should be our attitude to a result that is empirically ridiculous on its face but is a logical consequence of a set of axioms? Either we discover why the result is not really ridiculous, or we question the appropriateness of the axioms. To be content, for decades, to proceed with a formal result that still seems counterintuitive, simply on the basis of mathematical formalism, is not a very good idea. Although the Stein result has (happily) had virtually no impact on the way analysts in the real world make their estimates, over 400 papers have appeared in the statistical literature making further extensions of the James–Stein paper. For example, Lindley in commenting on a paper by Efron and Morris (1973) informs us:

> Now comes the crunch—notice it applies to the general linear model. The usual theory says x_i (maximum likelihood) is the best estimate of μ_i, but Stein showed that there is another estimate which is, for every set of μ's, better than it, when judged by the squared-error criterion except when only one or two parameters are involved. In other words, using standard criteria, the usual estimate is unsound. Further calculation (described in the paper) shows that it can be seriously unsound: with 10 parameters, quite a small number by the standard of present day applications, the usual estimate can have five times the squared error of Stein's estimate. And remember—it can never have smaller squared error ... the result of Stein undermines the most important practical technique in statistics ....
>
> The next time you do an analysis of variance or fit a regression surface (a line is all right!) remember you are for sure, using an unsound procedure ....
>
> Worse is to follow, for much of multivariate work is based on the assumption of a normal distribution. With known dispersion matrix this can again be transformed to the standard situation and consequently, in all cases except the bivariate one, the usual estimates of the means of a multivariate normal distribution are suspect ....

To get a better feel for what is happening, let us consider the one-dimensional case. Suppose we wish to estimate the mean of a random variable X on the basis of one observation of that random variable using estimators of the form

$$\mu_0 = aX. \tag{5.2.2}$$

We shall pick a in such a way as to minimize

$$Q(aX) = E[(aX - \mu)]^2 = a^2\sigma^2 + \mu^2(1 - a)^2. \tag{5.2.3}$$

Taking the derivative with respect to a and setting it equal to 0, we find the optimal a to be given simply by

$$a = \frac{\mu^2}{\mu^2 + \sigma^2}. \tag{5.2.4}$$

Using this a, we find that

$$Q(aX) = \frac{\mu^2}{\mu^2 + \sigma^2}\sigma^2 < \sigma^2. \tag{5.2.5}$$

Of course, in practice, we do not have μ or σ^2 available for our shrinkage factor a. Still, we should ask why it is that such a factor, were it realistically available, helps us. Perhaps we get some feel if we rewrite aX:

$$aX = \frac{X/\mu}{1 + (\sigma^2/\mu^2)}\mu. \tag{5.2.6}$$

This gives us the truth—μ—degraded by a multiplier which, if μ is small (relative to σ^2), would discount automatically large values of X as outliers. If μ is large (relative to σ^2), then we are left essentially with the usual estimator X. Thus, there is no paradox in the improvement of aX over X as an estimator for the one-dimensional case, if we know μ and σ^2. Note, moreover, that the argument to find a did not depend on any assumption of normality, only on the existence of a finite variance.

Again, in the one-dimensional case, we should address ourselves to dealing with the situation where we do not have μ or σ^2 available for our shrinkage factor. (I shall assume we do know σ^2 for reasons of convenience, but the argument holds if we do not have σ^2 and must estimate it.) In such a case, we have available the estimator $X^2/(X^2 + \sigma^2)X$. But here, we generally lose our "free lunch." If the data are normally distributed, then our risk curve looks like the one in Figure 5.2.2.

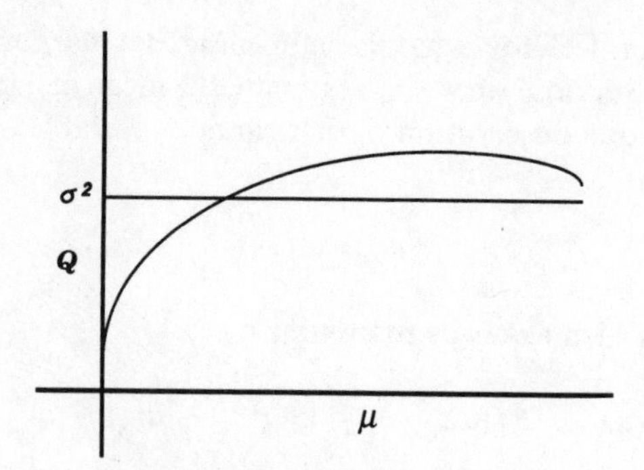

Figure 5.2.2. Risk of $[X^2/(X^2 + \sigma^2)]X$.

Let us suppose that we were allowed to use the following strategy (Thompson 1969): we have one observation X_1 to use for the estimation of μ. But, in addition, we have $k - 1$ additional observations $X_2, X_3, \ldots, X_k$ to be used in an a of the form

$$a = \frac{\sum_{i=1}^{k} X_i^2}{\left(\sum_{i=1}^{k} X_i^2 + k\sigma^2\right)}. \tag{5.2.7}$$

Now we know that the expectation of the square of a random variable X_i, with finite mean and variance, is given by

$$E(X^2) = \mu^2 + \sigma^2. \tag{5.2.8}$$

As we indicate in Appendix A.1, the average of samples of such a random variable will converge almost surely to its expectation; that is,

$$\sum_{i=1}^{k} \frac{X_i^2}{k} \to \mu^2 + \sigma^2, \quad \text{almost surely in } k. \tag{5.2.9}$$

Thus, our shrinkage factor a approaches

$$\frac{\mu^2 + \sigma^2}{\mu^2 + 2\sigma^2}, \quad \text{for } k \text{ large.} \tag{5.2.10}$$

This would give, for large k, X_1 and a being asymptotically independent,

$$Q(aX_1) \approx \frac{\mu^4 + 3\sigma^2\mu^2 + \sigma^4}{\mu^4 + 4\sigma^2\mu^2 + 4\sigma^4}\sigma^2 < \sigma^2. \tag{5.2.11}$$

We might suspect that something in the Stein formulation may allow such a phenomenon to occur. Indeed, this is the case. For the loss function considered

$$L(\mu^*, \mu) = \sum_{i=1}^{k} \frac{(\mu_j^* - \mu_j)^2}{k}, \tag{5.2.12}$$

and estimators of the form

$$\mu^* = \frac{X^T X}{X^T X + c\sigma^2} X, \tag{5.2.13}$$

it is straightforward to show that the risk is not dependent on the allocation of the $\{\mu_j\}$ for any fixed $\sum_j \mu_j^2$. (Alam and Thompson (1968) proved that, in the normal case, this estimator beats X for $p > 2$ if $0 < c < 2(p-2)$.) Accordingly, we need only consider the case where $\boldsymbol{\mu} = (\mu, \mu, \ldots, \mu)$. But this reduces immediately to the kind of one-dimensional estimator we showed had asymptotically (in p) smaller risk than X_1. (Apparently, for the normal case, the asymptotic result starts impacting for $p = 3$.) Thus, it is the assumption of a loss function of a *particular form* which gives the apparent Stein improvement.

Note that for unequal weights and unknown variance, the Stein result holds, *if we know the weights in the loss function*

$$L(\mu^*, \mu) = \sum_{j=1}^{k} \frac{w_j(\mu_j^* - \mu_j)^2}{k}. \tag{5.2.14}$$

But is it not reasonable to assume that we shall frequently know the weights precisely? After all, cost functions are frequently common. So, for example, we might need to estimate $\sum_j w_j \mu_j$, where the weights are known. Note that this is the one-dimensional estimation problem where we know, in the normal case, we cannot uniformly beat $\sum_j w_j X_j$.

The cases where we know the weights in the loss function in (5.2.14) are rare. Any strategy that assumes we do have precise knowledge of the weights is likely to be dangerous. Let us look at the more realistic situation where we do not have precise knowledge of the weights. Thus, let us consider the loss function

$$L(\mu^*, \mu) = \sum_{j=1}^{k} \frac{w_j^t(\mu_j^* - \mu_j)^2}{k}, \tag{5.2.15}$$

where, for all $t \in T$,

$$\sum_{j=1}^{k} w_j^t = 1, w_j^t \geq 0.$$

Let the risk be given by

$$Q(\mu^*; \mu, t) = E[L(\mu^*, \mu, t)]. \tag{5.2.16}$$

Let the class of estimators Δ to be considered be those of the form

(5.2.17) $\mu^* = Xf(X'X)$, where f is positive, real valued, and ≤ 1.

Definition. An estimator μ^* is said to be *w-admissible* if there does not exist in Δ an estimator μ^{**} such that $Q(\mu^{**}) \leq Q(\mu^*)$ for all (μ, t) and for at least one (μ, t), $Q(\mu^{**})$ is strictly less than $Q(\mu^*)$.

Definition. An estimator is *w-minimax* if it minimizes $\sup_{(\mu,t)} \{Q(\mu^*; \mu, t)\}$ for all members of Δ.

Note that the usual estimator $(X_1, X_2, X_3, \ldots, X_k)$ is w-admissible (consider the special case where $w_1^t = 1$). Moreover, $(X_1, X_2, X_3, \ldots, X_k)$ minimizes $\text{Max}_{\mu,t}\{Q\}$, that is, is *w-minimax*. The Stein estimators cannot be *w-minimax* for squared loss function, since for $w_1^t = 1$, they are randomized estimates of μ_1.

In conclusion, there is no "paradox" about Stein estimation. The free lunch is due to an apparent but artificial transferral of information between the dimensions as a result of an unrealistic assumption about the loss function. Shrinkage toward an arbitrary point (without prior information), on the basis of a factor that is built up using information from variables that are totally unrelated, which strikes most people at first glance as inappropriate, is indeed inappropriate.

When estimating, simultaneously, the density of mosquitoes in Houston, the average equatorial temperature of Mars, and the gross national product of ancient Persia, we ought not believe that some mathematical quirk demands that we multiply our usual (separable) estimates by a finagle factor which artificially combines all three estimates.

The above study has been given as an example of the difficulties which attend us when we attempt to make the world conform to an idealized mathematical construction, instead of the other way round. When the use of a particular criterion function yields results that are completely contrary to our intuitions, we should question the criterion function before disregarding our intuitions. At the end of the day, we may find that our intuitions were indeed wrong. The world is not flat, naive perceptions notwithstanding. However, the flatness of the earth was not disproved by construction of an artificial mathematical model, but rather by the construction of a model that explained real things with which the assumption of a flat earth could not cope.

References

Alam, Khursheed and Thompson, James R. (1968). Estimation of the mean of a multivariate normal distribution, Indiana University Technical Report.

Efron, Bradley and Morris, Carl (1973). Combining possibly related estimation problems, *JRSS B*, **35**, 379–421.

James, W. and Stein, Charles (1961). Estimation with quadratic loss function, *Proceedings of the Fourth Berkeley Symposium*, University of California Press, Berkeley, pp. 361–370.

Judge, George G. and Bock, M. E. (1978). *The Statistical Implications of Pre-Test and Stein Rule Estimators in Econometrics*, North Holland, New York.

Lehman, E. L. (1949). *Notes on the Theory of Estimation*, University of California Press, Berkeley.

Thompson, James R. (1968). Some shrinkage techniques for estimating the mean, *JASA*, **63**, 113–122.

Thompson, James R. (1968). Accuracy borrowing in the estimation of the mean by shrinkage to an interval, *JASA*, **63**, 953–963.

Thompson, James R. (1969). On the inadmissibility of X as the estimate of the mean of a p-dimensional normal distribution for $p \geq 3$, Indiana University Technical Report.

5.3. FUZZY SET THEORY

Generalization is a subject that justly concerns mathematicians. If a useful concept can be extensively generalized, then there is reason for excitement. We shall briefly outline below Lofti Zadeh's important generalization of the traditional binary logic system.

In the musical, *The King and I*, at one point the King of Siam delivers the following monologue: "When I was a boy, world was but a dot. What was so, was so. What was not, was not. Now I am a man. World has changed a lot. Some things nearly so, others nearly not."

The notion that things are nearly true or nearly false is not unlike the theory of "fuzzy set theory" developed and popularized by Lofti Zadeh. As Zadeh writes, "More often than not, the classes of objects encountered in the real physical world do not have precisely defined criteria of membership." In order to deal with this perceived phenomenon, Zadeh has modified the usual notions of naive set theory. For example, symbolically writing down a well understood (at least since the time of the classical Greek philosophers) concept that if A denotes a particular set, and x an entity at a lower hierarchy, either

$$x \in A \tag{5.3.1}$$

or

$$x \notin A. \tag{5.3.2}$$

Thus, a cow is either an animal or it is not.

Zadeh attempts to generalize such concepts by introducing a "membership function" $f_A(\cdot)$ which maps x to a number between 0 and 1. If $f_A(x)$ is close to 1 then x has a high degree of membership in A. It is "nearly so" that x is a member of A. On the other hand, if $f_A(x)$ is close to 0, then it is "nearly not so" that x is a member of A. One example given is the "fuzzy concept" of the set of numbers A much greater than 1. Is 1.1 much greater than 1? Is 2? Is 10^{12}? To many, such questions absent a meaningful context are simply meaningless. If we ask the question in terms of the load on a bridge relative to its recommended safe load, 1.1 is not a large number, 2 (using standard construction codes in the United States) is beginning to be. While 10^{12} is very large. Naturally, if it turns out that that the numbers presented were actually the true ratio raised to the 10^{20} power, then none of the numbers is "large" relative to 1. However, fuzzy set theory attempts to give us a means of answering such questions even absent context.

The notion of multivalued logics is not new with Zadeh. It was prevalent among some pre-Socratic Greek philosophers and is, at the informal level, operative among non-Western societies generally. In a very real sense, a key marker of the West is its reliance on the binary logic which characterizes rational thinking. As an example of a multivalued logic, we might say that a number between 1 and 10 was not much larger than 1; a number greater than 100 was much larger than 1; and numbers between 10 and 100 are neither much larger than 1 nor not much larger than 1. The concept of fuzzy set theory generalizes the membership notion from three levels to a continuum of levels. Just as traditional set theory can be regarded, perhaps, as a mathematization of the logic of Aristotle, just so fuzzy set theory might be regarded as a mathematization of the logic of the pre-Socratic Greek philosophers, who were comfortable with multivalued logics. In another sense, it can be viewed as an essential mathematization of the concepts of such modern "nonfoundational" philosophers as Jacques Derrida or Richard Rorty. Let us consider the Zadeh version of the usual set theoretic operations. For example

(5.3.3) Two fuzzy sets A and B are *equal* if and only if $f_A(x) = f_B(x)$ for all possible x.

(5.3.4) The *complement* A' of the set A is defined by the membership function $f_{A'}(x) = 1 - f_A(x)$.

(5.3.5) A is a *subset* of B if and only if for all x, $f_A(x) \leq f_B(x)$.

(5.3.6) The *union* of two fuzzy sets A and B is defined via the membership function $f_{A \cup B}(x) = \text{Max}[f_A(x), f_B(x)]$.

(5.3.7) The *intersection* of two fuzzy sets A and B is defined via the membership function $f_{A \cap B}(x) = \text{Min}[f_A(x), f_B(x)]$.

In order better to understand fuzzy set theory, let us use it to make a decision as to which place is better to live: Yalta or Houston. We have to list our criteria specifically if we are to use fuzzy set theory. So we define certain sets as being oriented toward the description of a favorable place:

(5.3.8)
$A =$ the set of places with a high material standard of living;
$B =$ the set of places with a high degree of personal freedom;
$C =$ the set of places with a low rate of crime;
$D =$ the set of places with a good winter climate.
$E =$ the set of places with a good summer climate.

Shall we then use as our criterion for a city being a good place to live the union of the five criteria above? Clearly not. Devil's Island was reputed to enjoy an excellent climate, but the inhabitants enjoyed little freedom, a high likelihood of being murdered by a convict, and a material standard of living close to starvation.

The natural criterion, using fuzzy set theory, is $f_{A \cap B \cap C \cap D \cap E}(\cdot)$ as the natural membership function for "good places to live." Suppose then the memberships of Yalta and Houston are given as below:

$$
\begin{array}{ll}
f_A(\text{Yalta}) = .3 & f_A(\text{Houston}) = .8 \\
f_B(\text{Yalta}) = .3 & f_B(\text{Houston}) = .9 \\
f_C(\text{Yalta}) = .95 & f_C(\text{Houston}) = .2 \\
f_D(\text{Yalta}) = .9 & f_D(\text{Houston}) = .9 \\
f_E(\text{Yalta}) = .9 & f_E(\text{Houston}) = .2.
\end{array}
\tag{5.3.9}
$$

We observe that

(5.3.10)
$$f_{A\cap B\cap C\cap D\cap E}(\text{Yalta}) = \text{Min}[.3,.3,.95,.9,.9] = .3;$$
$$f_{A\cap B\cap C\cap D\cap E}(\text{Houston}) = \text{Min}[.8,.9,.2,.9,.2] = .2.$$

Since $f_{A\cap B\cap C\cap D\cap E}(\text{Yalta}) > f_{A\cap B\cap C\cap D\cap E}(\text{Houston})$, the membership of Yalta in the set of desirable places to live is greater than that of Houston.

Let us now examine the situation when we define the set of bad places to live as those that do not do well with regard to $A \cap B \cap C \cap D \cap E$; that is, let us examine $f_{(A\cap B\cap C\cap D\cap E)'}(\cdot) = 1 - f_{(A\cap B\cap C\cap D\cap E)}(\cdot)$. We note that

(5.3.11)
$$f_{(A\cap B\cap C\cap D\cap E)'}(\text{Yalta}) = 1 - .3 = .7 < f_{(A\cap B\cap C\cap D\cap E)'}(\text{Houston}) = 1 - .2 = .8.$$

Or, alternatively, let us consider

(5.3.12)
$$f_{A'}(\text{Yalta}) = 1 - .3 = .7 \qquad f_{A'}(\text{Houston}) = 1 - .8 = .2$$
$$f_{B'}(\text{Yalta}) = 1 - .3 = .7 \qquad f_{B'}(\text{Houston}) = 1 - .9 = .1$$
$$f_{C'}(\text{Yalta}) = 1 - .95 = .05 \qquad f_{C'}(\text{Houston}) = 1 - .2 = .8$$
$$f_{D'}(\text{Yalta}) = 1 - .9 = .1 \qquad f_{D'}(\text{Houston}) = 1 - .9 = .1$$
$$f_{E'}(\text{Yalta}) = 1 - .9 = .1 \qquad f_{E'}(\text{Houston}) = 1 - .2 = .8.$$

The appropriate "set of badness" is clearly $A' \cup B' \cup C' \cup D' \cup E'$. So, quite consistently with customary considerations in naive set theory (De Morgan's law), we see that $f_{A'\cup B'\cup C'\cup D'\cup E'}(\cdot) = f_{(A\cap B\cap C\cap D\cap E)'}(\cdot)$. Other notions proceed quite intuitively. For example, we can say that *A is smaller than or equal to B if and only if* $f_A(x) \le f_B(x)$ *for all* x.

However, the above example shows one practical difficulty with fuzzy set theory: quite reasonable membership functions can result in answers that do not jibe with the kinds of intuitive answers at which most would arrive using more standard analyses. Although Yalta is one of the most affluent and pleasant resorts in the Soviet Union, it probably would not be the location of choice for most educated people when compared with Houston (in spite of the high crime rate and miserable summers of the latter city). In carrying out our fuzzy set theory analysis, we essentially carried out a maximin strategy: we based our decision on maximizing the worst performance indicator. The psychological fact is that people simply do not usually make decisions this way. They would be much more likely to put weightings on standard of living and on personal freedom that were many times those placed on crime rate or climate. Furthermore, they would probably put certain restrictions concerning one or more criteria on the kinds of places they would not live regardless of performance on all other criteria. So one very practical problem with fuzzy set theory is that it leads to decisions very different from those reached intuitively.

There is a misconception among some that fuzzy set theory is simply Bayesian analysis by another name. To illustrate the rather fundamental difference between the two notions, let us investigate the probability that Iowa will, next year, enjoy successful wheat production. Let A be the class of years of successful wheat production. The Bayesian analysis would start off with the historical record and come up with an initial prior probability that next year will be successful from the standpoint of wheat production, say $P(A)$. As time proceeds, we accumulate a certain amount of relevant meteorological data, say H_1, which can be used to give us an updated (posterior) estimate, $P(A|H_1)$. As more and more information comes in, as we can see the young wheat crop and then the more mature crop from satellite, and as we get reports from farmers and county agents, we can update our estimate further to obtain $P(A|H_1, H_2, \ldots, H_n)$. After we get the final bit of information, which will consist of harvest figures, we generally are able to come up with a probability of one or a probability of zero.

Fuzzy set theory is not directly concerned with this kind of analysis (though a Bayesian analysis can be performed in a fuzzy set theoretic situation). Rather, it is concerned with defining the membership function $f_A(\cdot)$. What does it take to have a very successful corn year? Surely, if x were a production of 100,000,000 metric tons, we would wish to say that f_A was very close to 1. If, on the other hand, x were 0, we would say that f_A was close to 0. Any kind of reasonable analysis would have gradations of successful corn production. But it is not at all clear that, in assessing a number of "quality of life" factors for citizens of Iowa by the use of fuzzy set theory, one comes up with criteria that have much relationship with reality. Also, a market analyst is unlikely to be very content with trying to pool "successful" productions of corn from Iowa, Nebraska, Kansas, and so on. Rather, the analyst will try to make numerical estimates of their respective productions so that some comparisons can be made between anticipated supply and demand. The analyst, quite properly, will be inclined to undertake something like a time-evolving Bayesian analysis of the posterior probability that the productions will lie in certain intervals. As the harvest comes in, an efficient market will determine the prices rather unambiguously. Most likely, the market analyst would tend to regard a fuzzy set theoretic approach as being unnecessarily subjective, nonquantitative, and not necessarily very responsive to facts.

On the other hand, a director of Sovkhozes who was carrying out a corn production analysis in the Ukraine might find a fuzzy set theoretic approach personally, if not agriculturally, useful. The ability to define the membership function (as it essentially must be) subjectively and to revise the function as time progresses gives the director an opportunity to improve his success record irrespective of the actual production levels *and without any necessity of data falsification.* A kind of "who controls the present controls the past and who controls the past controls the future" situation.

James Burnham, the philosopher whose work *The Managerial Revolution* inspired, in large measure, Orwell to write *1984,* defined "dialectical materialism" as a pseudological means by which reality can be made to conform to theory. Certainly, there is some danger that fuzzy set theory can become a device that is tailor-made for such a purpose. Burnham noted that as capitalist corporations became large and inefficient, their directors would also tend to adopt arguments that would enable an appearance of excellence in spite of unfavorable bottom lines.

Interestingly, although fuzzy set theory was first developed at the University of California, it does not yet have much popularity in the United States or the other Western democracies. It is extremely popular in some countries in eastern Europe, particularly in Roumania. It is popular in Japan (even with some major companies), which places rationalism below such notions as concordance. In the United States, it has some popularity with several corporations and in some schools of business.

We started, in the Introduction of this book, with Winston Smith's famous "Freedom is the freedom to say that two plus two make four. Given that, all else follows." It is clear that fuzzy set theory can be used, quite easily, to extend this freedom to those who want to say that two plus two make five. It is, perhaps, the most intrinsically antirationalist concept to have surfaced in modern science, for it is an assault on logic itself. Needless to say, in the decades since its introduction, fuzzy set theory has not proved to be a useful tool in the design of airplanes or computers. In a society that is increasingly willing to avoid distinctions and to avoid bottom lines, however, it appears likely to become ever more important as a device for the management of people and institutions.

References

Burnham, James (1941). *The Managerial Revolution,* Day, New York.

Orwell, George (1949). *1984,* Harcourt, Brace & Company, New York.

Zadeh, L. A. (1965). Fuzzy sets, *Information and Control,* **8,** 338–353.

5.4. QUALITY CONTROL

The science of quality control is certainly not a false trail; it is a study which any modern organization disregards at its peril. On the other hand, there are a number of paradoxes in quality control. A "paradoxical" subject, of course, ceases to be a paradox once it is correctly perceived. It is not so much the purpose of this section to go into great detail concerning control charts and other technical details of the subject as it is to uncover the essence of the evolutionary optimization philosophy, which is always the basis for an effective implementation of quality control. Most of us, including many quality control professionals, regard quality control as a kind of policing function. The quality control professional is perceived as a find of damage control officer; he or she tries to keep the quality of a product controlled in the range of market acceptability. The evolutionary optimization function of quality control is frequently overlooked.

Imagine a room filled with blindfolded people which we would wish to be quiet but is not because of the presence of a number of noise sources. Most of the people in the room are sitting quietly and contribute only the sounds of their breathing to the noisiness of the room. One individual, however, is firing a machine gun filled with blanks, another is playing a portable radio at full blast, still another is shouting across the room, and, finally, one individual is whispering to the person next to him.

Assume that the "director of noise diminution" is blindfolded also. Any attempt to arrange for a quiet room by asking everyone in the room to cut down his noise level 20% would, of course, be ridiculous. The vast majority of the people in the room, who are not engaged in any of the four noise making activities listed, will be annoyed to hear that their breathing noises must be cut 20%. They rightly and intuitively perceive that such a step is unlikely to do any measurable good. Each of the noise sources listed is so much louder than the next down the list that we could not hope to hear, for example, the person shouting until the firing of blanks had stopped and the radio had been turned off.

The prudent noise diminution course is to attack the problems sequentially. We first get the person firing the blanks to cease. Then, we are able to hear the loud radio, which we arrange to have cut off. Then we can hear the shouter and request that he be quiet. Finally, we can hear the whisperer and request that he also stop making noise.

If we finally have some extraordinary demands for silence, we could begin to seek the breather with the most clogged nasal passages, and so on. But generally speaking, we would arrive, sooner or later, at some level of silence which would be acceptable for our purposes.

This intuitively obvious analogy is a simple example of the key notion of quality control. By standards of human psychology, the example is also rather bizarre. Of the noise making individuals, at least two would be deemed sociopathic. We are familiar with the fact that in most gatherings, there will be a kind of uniform buzz. If there is a desire of a master of ceremonies to

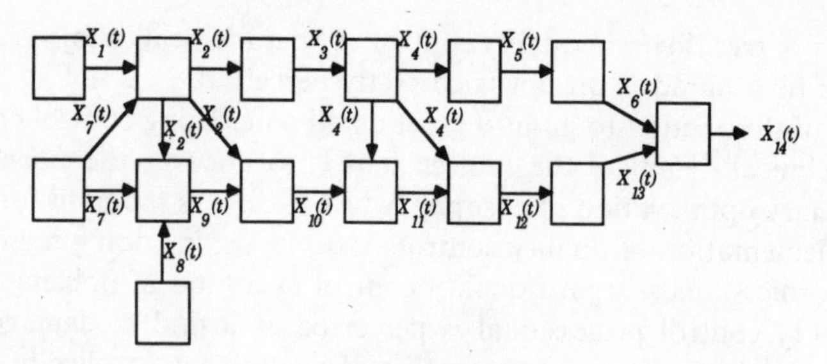

Figure 5.4.1

quieten the audience, it is perfectly reasonable for him to ask everyone please to be quiet. The fact is that machines and other systems tend to function like the (by human standards) bizarre example and seldom behave like a crowd of civilized human beings. It is the tendency to anthropomorphize systems that makes quality control, correctly perceived, appear paradoxical.

The above is an example of an empirical observation of the Italian sociologist Vilfredo Pareto: *a relatively few failure reasons are responsible for the catastrophically many failures.*

Let us illustrate this concept in the network diagram in Figure 5.4.1. Each one of the boxes represents some modular task in a production process. It is in the nature of the manufacturing process that it is desirable that the end product output $X_{14}(t)$ be maintained as constant as possible. For example, a company that is making a particular type of machine bolt will want to have them all the same, for the potential purchasers of the bolt are counting on a particular diameter, length, and so on. It is not unheard of for people to pay thousands of dollars for having a portrait painted from a simple photograph. The artist's ability to capture and embellish some aspect he perceives in the photograph is (quite rightly) highly prized. A second artist would produce, from the same photograph, quite a different portrait. No one would like to see such subjective expression in the production of bolts. If we allowed for such variability, there would be no automobiles, no lathing machines, and no computers. This fact does not negate aesthetic values. Many of the great industrial innovators have also been major patrons of the arts. Modularity demands uniformity. This fact does not diminish the creative force of those who work with modular processes any more than a net interferes with the brilliance of a tennis professional. And few workers in quality control would wish to have poems written by a CRAY computer.

Departures from uniformity, in a setting where uniformity is desired, furnish a natural means of evolutionary improvement in the process. Now by Pareto's maxim, if we see variability in $X_{14}(t)$, then we should not expect to find the source of that variability uniformly distributed in the modules of the process

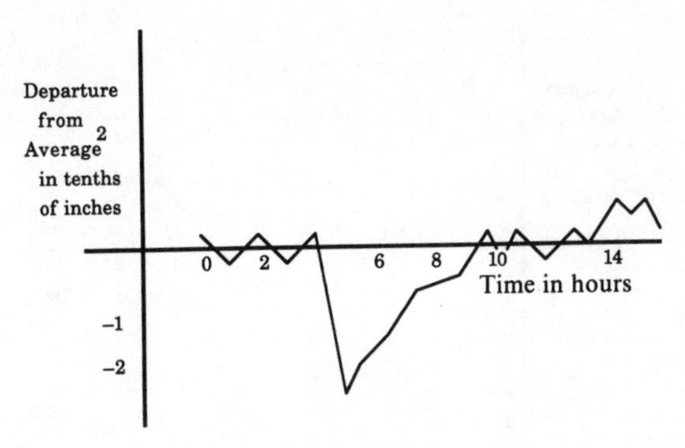

Figure 5.4.2

upstream. Generally, we shall find that one of the modules is offering a variability analogous to that of the machine gun firing blanks in an auditorium. The fact that this variability is very likely intermittent in time offers the quality control investigator a ready clue as to where the problem is. Naturally, there may be significant time lags between $X_3(t)$, say, and $X_{14}(t)$. These are usually well known; for example, we usually know how far back the engine block is installed before final inspection takes place. Thus, if a particular anomaly in the final inspection is observed during a certain time epoch, the prudent quality control worker simply tracks back the process in time and notes variabilities in the modules which could impact on the anomaly in the proper time frame. Although they are useful for this purpose, statistically derived control charts are not absolutely essential. Most of the glitches of the sort demonstrated in Figure 5.4.2 are readily seen with the naked eye. The Model T Ford, which had roughly 5000 parts, was successfully monitored without sophisticated statistical charts. We show then the primitive *run chart* in Figure 5.4.2.

Once we have found the difficulty that caused the rather substantial glitch between hours 4 and 8, we have significantly improved the product, but we need not rest on our laurels. As time proceeds, we continue to observe the run charts.

We note a similar kind of profile in Figure 5.4.3 to that in Figure 5.4.2. However, note that the deviational scale has been refined from tenths to hundredths of an inch. Having solved the problem of the machine gun firing blanks, we can now approach that of the loud radio. The detective process goes forward smoothly (albeit slowly) in time. Ultimately, we can produce items that conform to a very high standard of tolerance.

A second paradox in quality control, again due to our tendency to treat machines and other systems as though they were human beings, has to do

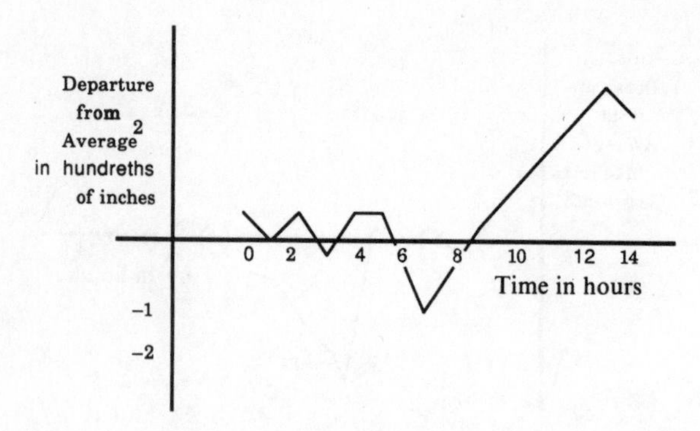

Figure 5.4.3

with the false perception that a source of variability, once eliminated, is very likely to occur again as soon as we turn our attention to something else. In many human activities, there is something like an analogy to a man juggling balls. As he hurls a red ball upward, he notices that he must immediately turn his attention to the yellow ball which is falling. When he hurls the yellow ball upward, he notes that the green ball is falling. By the time he throws the green ball upward, he must deal with the red ball again. And so on. If a human decides to give up smoking, he is likely to note an increase in his weight. Once he worries about weight control, he may start smoking again.

Happily, machines suffer from no such difficulty. The author drives a 10 year old Buick. The engine has experienced approximately 1 billion revolutions and still functions well. No one is surprised at such reliability with mechanical devices. The ability to execute identical operations many times without fail is in the very nature of machines. Interestingly, there are many systems in which similar reliability is to be expected. In highly automated industrial systems, once a source of variability has been eliminated, it is unlikely to be a problem again unless we are very careless indeed. The V-8 engine, perfected over 30 years ago, still functions very well. A newer, more sophisticated engine may be much less reliable unless and until we have carefully gone through the quality control optimization process for it as well. Innovation is usually matched by new problems of variability not previously experienced in older systems. This does not speak against innovation; rather it warns us that we simply do not reasonably expect to have instantaneous quality control with a new system. Companies and other organizations that expect to go immediately from a good idea to a successful product are doomed to be disappointed. This was well understood by the founder of the assembly line, Eli Whitney, who started the modern industrial production process in 1798 with his musket factory. Whitney's product involved the construction and assembly of approximately 50 basic parts. The Model T Ford, developed roughly 100 years later, involved some 5000 basic parts. The basic paradigms of both Whitney ad Ford were essentially the same:

1. Eliminate most potential problems at the design stage.
2. Use extensive pilot study testing to eliminate remaining undiscovered problems.
3. Use testing at the production stage to eliminate remaining glitches as a means of perfecting the product, always remembering that such glitches are generally due to defects in a few submodules of the production process rather than general malaise.

Henry Ford (1926) attempted to codify his ideas in his four principles of manufacturing:

1. An absence of fear of the future or veneration of the past. One who fears failure limits his activities. Failure is only the opportunity to begin again. There is no disgrace in honest failure; there is disgrace in fearing to fail. What is past is useful only as it suggests ways and means for progress.
2. A disregard of competition. Whoever does a thing best ought to be the one to do it. It is criminal to try to get business away from another man—criminal because one is then trying to lower for personal gain the condition of one's fellowmen—to rule by force instead of by intelligence.
3. The putting of service before profit. Without a profit, business cannot expand. There is nothing inherently wrong about making a profit. Well-conducted business enterprise cannot fail to return a profit, but profit must and inevitably will come as a reward for good service—it must be the result of service.
4. Manufacturing is not buying low and selling high. It is the process of buying materials fairly and, with the smallest possible addition of cost, transforming those materials into a consumable product and giving it to the consumer. Gambling, speculating, and sharp dealing tend only to clog this progression.

Ford's four principles no longer sound very modern. In the context of what we are accustomed to hear from America's contemporary captains of industry, Ford's four principles sound not only square but rather bizarre. That is unfortunate, for I do not believe that they would sound bizarre to a contemporary Japanese, Taiwanese, Korean, or West German industrialist. The modern paradigm of quality control is perhaps best summarized in the now famous 14 points of W. E. Deming (1982), who is generally regarded as the American apostle of quality control to Japan:

1. Create constancy of purpose toward improvement of product and service, with a plan to become competitive and to stay in business. Decide to whom top management is responsible.

2. Adopt the new philosophy. We are in a new economic age. We can no longer live with commonly accepted levels of delays, mistakes, defective materials, and defective workmanship.
3. Cease dependence on mass inspection. Require instead statistical evidence that quality is built in, to eliminate need for inspection on a mass basis. Purchasing managers have a new job and must learn it.
4. End the practice of awarding business on the basis of price tag. Instead, depend on meaningful measures of quality, along with price. Eliminate suppliers that cannot qualify with statistical evidence of quality.
5. Find problems. It is management's job to work continually on the system (design, incoming materials, composition of material, maintenance, improvement of machine, training, supervision, retraining).
6. Institute modern methods of on-the-job training.
7. Institute modern methods of supervision of production workers. The responsibility of foremen must be changed from sheer numbers to quality. Improvement of quality will automatically improve productivity. Management must prepare to take immediate action on reports from foremen concerning barriers such as inherited defects, machines not maintained, poor tools, or fuzzy operational definitions.
8. Drive out fear, so that everyone may work effectively for the company.
9. Break down barriers between departments. People in research, design, sales, and production must work as a team, to foresee problems of production that may be encountered with various materials and specifications.
10. Eliminate numerical goals, posters, and slogans for the work force, asking for new levels of productivity without providing methods.
11. Eliminate work standards that prescribe numerical quotas.
12. Remove barriers that stand between the hourly worker and his right to pride of workmanship.
13. Institute a vigorous program of eduction and training.
14. Create a structure in top management which will push every day on the above 13 points.

Problems in production are seldom solved by general broad spectrum exhortations for the workers to do better. An intelligent manager is more in the vein of Sherlock Holmes than that of Norman Vincent Peale. This rather basic fact has escaped the attention of most members of the industrial community, including practitioners of quality control. To give an example of this fact, we note the following quotation from the Soviet quality control expert Ya. Sorin (1963):

> Socialist competition is a powerful means of improving the quality of production. Labour unions should see to it that among the tasks voluntarily assumed by brigades and factories ... are included those having to do with improving quality

> control .... In the course of socialist competition new forms of collective fight for quality control are invented. One such form is a "complex" brigade created in the Gorky car factory and having as its task the removal of shortcomings in the design and production of cars. Such "complex brigades" consist of professionals and qualified workers dealing with various stages of production. The workers who are members of such brigades are not released from their basic quotas. All problems are discussed after work. Often members of such brigades stay in factories after work hours to decide collectively about pressing problems.

In the above, we note the typical sloganeering and boosterism which is the hallmark of a bad quality control philosophy. The emphasis, usually, of bad quality control is to increase production by stressing the workers. The workers are to solve all problems somehow by appropriate attitudinal adjustment. Quality control so perceived is that of a man juggling balls and is doomed to failure.

Of course, Soviet means of production are proverbial for being bad. Surely, such misconceptions are not a part of American quality control? Unfortunately, they are. Let us consider an excerpt from a somewhat definitive publication of the American Management Association *Zero Defects: Doing It Right the First Time* (1965, pp. 3–9):

> North American's PRIDE Program is a positive approach to quality that carries beyond the usual system of inspection. It places the responsibility for perfection on the employee concerned with the correctness of his work. Its success depends on the attitude of the individual and his ability to change it. This program calls for a positive approach to employee thingking .... A good promotional campaign will help stimulate interest in the program and will maintain interest as the program progresses. North American began with a "teaser" campaign which was thoughtfully conceived and carried out to whet the interest of employees prior to the kick-off day .... Various things will be needed for that day—banners, posters, speeches, and so forth .... Many of the 1,000 companies having Zero Defects programs give employees small badges or pins for signing the pledge cards and returning them to the program administrator. Usually these badges are well received and worn with pride by employees at all times on the job .... In addition, it is good to plan a series of reminders and interest-joggers and to use promotional techniques such as issuing gummed stickers for tool boxes, packages and briefcases. Coffee cup coasters with printed slogans or slogans printed on vending machine cups are good. Some companies have their letterhead inscribed with slogans or logs which can also be printed on packing cases.

There is little doubt that an intelligent worker confronted with the Zero Defects program could think of additional places where management could stick their badges, pins, and posters. A worker, particularly an American worker, does not need to be tricked into doing his or her best. In a decently managed company, employee motivation can be taken as a given. But a human being is not a flywheel. The strong point of humans is their intelligence, not their regularity. Machines, on the other hand, can generally be counted on for their regularity. The correct management position in quality control is to treat each human worker as an intelligent, albeit erratic, manager of one or more regular, albeit nonreasoning, subordinates (machines). Thus, human workers should be perceived of as managers, problem finders, and problem solvers. It is a management misunderstanding of the proper position of human workers in our high-tech society which puts American industry so much at the hazard.

The basic technique of statistical quality control is quite simple. We show from the derivation of the central limit theorem in Appendix A.1 that, under very general conditions, a sample of independent and identically distributed observations $\{x_1, x_2, \ldots, x_n\}$ from a distribution having mean μ and variance σ^2 has, for n large, a sample average that is approximately normally distributed with mean μ and variance σ^2/n. This means that we would expect to find a sample mean outside the interval $\bar{\bar{X}} \pm 3\sigma/\sqrt{n}$ approximately 1 time in 500. A sample mean outside the interval $\bar{\bar{X}} \pm 2\sigma/\sqrt{n}$ would occur approximately 5 times in 100. Accordingly, when a sample mean value fell outside the $\bar{\bar{X}} \pm 2\sigma/\sqrt{n}$ interval, we would begin to suspect that something had changed in the system (hence it might be "out of control"), and when the value fell outside the interval $\bar{\bar{X}} \pm 3\sigma/\sqrt{n}$, we would be reasonably certain the system was out of control.

As a practical matter, we may have samples so small that the central limit theorem does not apply. And, of course, we shall generally have to use the sample variance s^2 instead of σ^2. This will usually not disturb us unduly, since we are simply trying to obtain a rough quantitative rule for implementing the run chart in Figure 5.4.2. In Figure 5.4.4 we indicate such a *control chart*. The

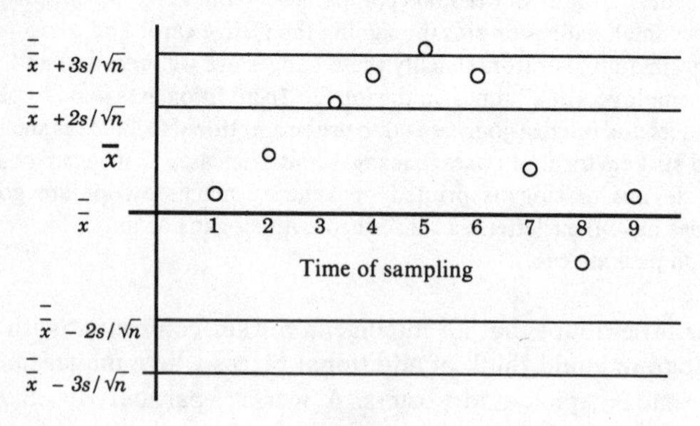

Figure 5.4.4

double barred sample mean indicates the average over an epoch of time which may be several days or weeks. n is the size of each sample. Similarly, the sample variance represents the average of the sample values over a number of samples.

In the above chart, we noted that something caused the system to go out of control around the time the fifth sample was collected. An examination revealed that the difficulty was caused by a vibration due to a faulty bearing in one of the machines. After the repair was effected, the system came back into control.

The above example is somewhat underwhelming as an example of evolutionary optimization. It seems to be simply maintenance. But suppose that management noted that such bearing wear-out was a commonly occurring event and decided to do something about it. Accordingly, they install an early warning detector for bearing vibration so that before any harm is done, the offending machine can be taken off line and repaired. Such a modest step could indeed be significant in the process of increasing uniformity of output.

A few years ago, one of the major manufacturers of American cars began to put Japanese transmissions on one of their lines. The introduction, however, was only partial in the sense that some cars in the line had Japanese transmissions, while some had American ones. Dealers in the area where the car was assembled were confronted with potential customers, obviously American workers in the assembly plant, who would ask a dealer whether the transmission was American or Japanese made. At first the dealers thought this was a kind of "buy American" intimidation on the part of these customers. However, they quickly noted that the workers were selectively purchasing the cars with Japanese transmissions. A report of this fact to the company caused a thorough testing of 40 American and 40 Japanese transmissions. All the parts in the American transmissions were within the company's specifications of "acceptable" variability. However, the Japanese transmissions were built to such close tolerances that deviations from the target values were simply not measurable using the calipers used by the American inspectors. The staff at the Japanese transmission plant had gone through the kind of evolutionary optimization which brings the product ever closer to the goal of uniformity.

In summary then, quality control is a bit of a misnomer. A better name would be "quality optimization."

References

Deming, W. Edwards (1982). *Quality, Productivity and Competitive Position,* Center for Advanced Engineering Studies, Boston, pp. 16–17.

Falcon, William, D. (ed.) (1965). *Zero Defects: Doing It Right the First Time,* American Management Association, Inc., New York.

Ford, Henry (1926). *My Life and Work,* Cornstalk Press, Syndey, p. 273.

Sorin, Ya. (1963). *On Quality and Reliability: Notes for Labor Union Activists,* Profizdat, Moscow, p. 120.

Thompson, James R. (1985). American quality control: What went wrong? What can we do to fix it? *Proceedings of the 1985 Conference on Applied Analysis in Aerospace, Industry and Medical Sciences,* Raj Chhikara, ed., University of Houston, Houston, pp. 247–255.

Todt, Howard C. (1965). Employee motivation: Fact or fiction, in *Zero Defects: Doing It Right the First Time,* William Falcon, ed., American Management Association, New York, pp. 3–9.

Appendix

A.1. A BRIEF PRIMER IN STOCHASTICS

Many processes in nature and in society are the macro results of the concatenation of a myriad of seemingly random events. For example, the temperature measured by a thermometer located at a point in a room is merely a measure of the average kinetic energy of millions of gas molecules striking the thermometer. In such a case, it may be natural to come up with some means of summing the microphenomena in such a way as to determine an average.

In other situations, it may appear that we are interested in the microphenomena themselves. For example, a poker player may wish to compute the probability that the dealing of four straight royal flushes to his opponent gives evidence that the dealer is cheating.

In dealing with stochastic phenomena, it seems most natural to begin with some variant of the classical arguments engaged in by the gentlemen amateur scholars of the 17th century, who were much concerned with games of chance. In a way, it is unfortunate that this path is the easiest to travel, for it may give the reader the false notion that probability theory is good for nothing more than dealing with artificial intellectual games far removed from reality.

A.1.a. Some Simple Combinatorics

Let us compute the number of ways that we can arrange in a distinctive order k objects selected without replacement from $(n \geq k)$ distinct objects. We note that there are n ways of selecting the first object, $n - 1$ ways of selecting the second object, and so on until we see that there are $n - k + 1$ ways of selecting the kth object. The total number of ways is called the $P(n, k)$, the *permutation* of n objects taken k at a time and is seen to be given by

$$P(n, k) = n(n-1)(n-2)\cdots(n-k+1) = \frac{n!}{(n-k)!}. \tag{A.1.1}$$

Next, we attempt to ask the question as to how many ways we can select k objects from n objects when we are not concerned with the distinctive order of selection. This is called the *combination* of n objects taken k at a time. We shall find this by noting that $P(n, k)$ could be first computed by finding $C(n, k)$ and then multiplying it by the number of ways k objects could be distinctly arranged (i.e., $k!$). So we have

$$P(n,k) = C(n,k)P(k,k) = C(n,k)k! \tag{A.1.2}$$

and thus

$$\binom{n}{k} = C(n,k) = \frac{P(n,k)}{k!} = \frac{n!}{(n-k)!k!}. \tag{A.1.3}$$

For example, the game of stud poker consists in the drawing of 5 cards from a 52 card deck (4 suits, 13 denominations). The number of possible hands is given by

$$C(52,5) = \frac{52!}{47!5!} = 2{,}598{,}960. \tag{A.1.4}$$

We are now in a position to compute some basic probabilities. Each of the 2,598,960 possible poker hands is equally likely (to see that this is so, think of the fact that if each of the cards is turned face down, we see only the back, which is the same for each card). To compute the probability of a particular hand, we simply evaluate

$$P(\text{hand}) = \frac{\text{number of ways of getting the hand}}{\text{number of possible hands}}. \tag{A.1.5}$$

Suppose we wish to find the probability of getting four of a kind (e.g., four aces, four kings). There are $C(13, 1)$ ways of picking a denomination, $C(4, 4)$ ways of selecting all the four cards of the same denomination, and $C(48, 1)$ of selecting the remaining card. Hence,

$$\begin{aligned} P(\text{four of a kind}) &= \frac{C(13,1)C(4,4)C(48,1)}{C(52,5)} \\ &= \frac{(13)(1)(48)}{2{,}598{,}960} = .00024. \end{aligned} \tag{A.1.6}$$

Similarly, to find the probability of getting two pairs, we have

(A.1.7)
$$P(\text{two pairs}) = \frac{C(13,2)C(4,2)C(4,2)C(44,1)}{C(52,5)} = \frac{(78)(6)(6)(44)}{2{,}598{,}960} = .0475.$$

A.1.b. Some Classical Probability Distributions

More generally, let us suppose we have an urn with k types of objects of numerosity $N_1, N_2, \ldots, N_k$. We wish to find the probability that, if we draw a total of $n_1 + n_2 + \cdots + n_k$ objects, then we have n_1 of the first type, n_2 of the second type, ..., n_k of the kth type. If we denote the k counters as $X_1, X_2, \ldots, X_k$, respectively, then we see that

(A.1.8)
$$P(X_1 = n_1, X_2 = n_2, \ldots, X_k = n_k) = \frac{C(N_1, n_1)C(N_2, n_2)\cdots C(N_k, n_k)}{C(N_1 + N_2 + \cdots + N_k, n_1 + n_2 + \cdots + n_k)}.$$

This is the *probability function* for the *multivariate hypergeometric* distribution. In some sense, because it provides an exact consequence of a set of axioms which fairly accurately describe many phenomena, the hypergeometric distribution is regarded as a *natural* (i.e., non ad hoc) distribution. It has been widely used by physicists and physical chemists for over 100 years (and by gamblers for 300 years). We note that this distribution is *discrete*; that is, the X_i's can take only values that are either integers or can be put into one-to-one correspondence with some subset of the integers. The $(X_1, X_2, \ldots, X_k)$ counters provide an example of a *random variable*, that is, a set of values in k-dimensional Euclidean space which correspond to events whose probability is well defined.

Dealing with multivariate phenomena is frequently quite complicated. Moreover, we live (at the naive level) in a three-dimensional space. Consequently, our intuitions are not very good in spaces beyond three. Furthermore, until the advent of computer graphics, we did not have an easy means of sketching curves in dimensions greater than two. When we sketch the probability function of a random variable, the probability function itself requires a dimension. Consequently, researchers generally try very hard to deal mainly with one-dimensional random variables. In reality, phenomena are usually multidimensional. So our own perceptual and (until the advent of the computer) computational limitations seem to compel us to look at low-dimensional projections even when we know that they tell only part of the story. We have shown, for example, in the section on SIMEST, how we can frequently overcome the computational requirements to deal with problems of low dimensionality.

For the moment, however, we shall follow tradition and look at the one-dimensional hypergeometric distribution. Here, there are $N - M$ balls of a type of interest, M balls of other type(s), and we wish to find the probability of getting x balls of the type of interest out of a draw, without replacement, of n balls from the totality of N balls. From (A.1.8), we immediately see that

$$P(X = x) = p(x) = \frac{C(N - M, x)C(M, n - x)}{C(N, n)}. \tag{A.1.9}$$

If n is large, we again run into a perceptual limitation (can't see the forest for the trees). So it is natural to attempt to summarize information about $p(x)$ with information of lower dimension than n. The most common of such summaries is the *mean*, or *expected value*. In the above example, this would simply be

$$E(X) = 0p(0) + 1p(1) + 2p(2) + \cdots + np(n). \tag{A.1.10}$$

$$\mu = E(X) = \sum xp(x). \tag{A.1.11}$$

Equation (A.1.11) is the more compact version of (A.1.10). For the hypergeometric distribution, we note that

$$\begin{aligned} E(X) &= \sum_{x=0}^{x=N} x\frac{C(N - M, x)C(M, n - x)}{C(N, n)} \\ &= \sum_{x=0}^{x-N} x\frac{(N - M)(N - M - 1)!}{(N - M - x)!x(x - 1)!}\frac{C(M, n - x)n(n - 1)!(N - n)!}{N(N - 1)!} \\ &= n\frac{N - M}{N}\sum_{x=0}^{x-N}\frac{C(N - 1 - M, x - 1)C(M, n - 1 - \{x - 1\})}{C(N - 1, n - 1)} \\ &= n\frac{N - M}{N}\sum_{y=0}^{y=S}\frac{C(S - M, y)C(M, r - y)}{C(S, y)}, \end{aligned} \tag{A.1.12}$$

where $S = N - 1$, $y = x - 1$, and $r = n - 1$. But the sum shown at the end is simply the sum over all possible values of a hypergeometric probability function with $N - M - 1$ balls of the desired type and M other balls when $n - 1$ draws are made. Since it covers all possibilities, this sum must be one. Thus, for the hypergeometric distribution,

$$E(X) = \mu = np, \quad \text{where } p = (N - M)/N. \tag{A.1.13}$$

The above represents one of the most common "tricks" in dealing with sums involving probabilities, namely, to manipulate the formula inside the sum until we arrive at something that sums to one. The expectation operator $E(\cdot)$ is seen to be simply an integration or summing operator with kernel $p(x)$. After a little familiarity, one will begin to deal with expectations as comfortably as with integrations.

We note that (A.1.13) makes intuitive sense. If a pollster goes out to sample 100 voters from a population in which 55% are Republicans, he will "expect" to get 55 in his sample. Naturally, he knows that for a particular sample, he will get precisely 55 only rarely. But if he carried out the sampling on another set of voters, then on another, and so on, he would expect that the overall average would be very nearly 55 Republicans per 100 voters. We note that the pollster could have disdained to deal with expectation and looked rather at the full representation of the probability function, that is, $\{p(0), p(1), p(2), \ldots, p(100)\}$, a 101-tuple. The one-dimensional expectation captures, in this example (and for all hypergeometric situations), all the information of the 101-tuple, for, if the pollster knows $\mu = n(N - M)/N$, he can readily find $(N - M)$, and thus M and thus every one of the terms in the 101-tuple, using (A.1.9).

For most stochastic situations, one summary number will not completely specify the system. It will not simply be enough to know μ. We might reasonably look at expected measures of departure from μ. For example, the *variance* is given by

$$E[(X - \mu)^2] = \mathrm{Var}(X) = \sigma^2 = \sum (X - \mu)^2 p(X). \tag{A.1.14}$$

For the hypergeometric distribution, an argument like that in (A.1.12) gives us

$$\begin{aligned} E[(X - \mu)^2] &= \sigma^2 = np(1 - p)(N - n)/(N - 1) \\ &= npq(N - n)/(N - 1). \end{aligned} \tag{A.1.15}$$

It is usually more natural to use the sampling without replacement distribution (i.e., the hypergeometric) than the sampling with replacement distribution (i.e., the *binomial*). However, for reasons of computational convenience, the binomial is frequently used as an approximation to the hypergeometric distribution.

To describe the binomial distribution, let us suppose that we have an urn filled with an infinite number of balls, the proportion of the balls of interest being given by p, the other balls being of proportion $q = 1 - p$. If we make a draw of size n from the urn, we would like to find the probability $p(x)$ that we obtain a number of balls of the type of interest equal to x. To do this, we note that one such possibility of drawing x balls of the type of interest is that the first x draws are of the type of interest and the next $n - x$ balls are balls of other type. Clearly, this draw has probability $pp \cdots p\, qq \cdots q = p^x q^{n-x}$. We note that the total number of ways we can get a configuration with x balls of one type selected from n is simply $C(n, x)$. Thus, the binomial probability function is given by

$$p(x) = C(n, x) p^x q^{n-x}. \tag{A.1.16}$$

An argument similar to that used in (A.1.12) shows that, for the binomial distribution, $\mu = np$ and $\sigma^2 = npq$.

In practice, it is clear that if n is small relative to N, the binomial approximation to the hypergeometric is good. Thus, a sample of 10,000 voters from the American electorate can be (and is) handled using the binomial distribution. Let us suppose that we wish to examine a sample of 10 voters from the national electorate. If we believe the proportion of Democrats is 40%, we wish to determine the probability that in our sample we find no more than three Democrats. This is readily seen to be given by

$$\begin{aligned} P(X \leq 3) &= C(10,0)(0.4)^0(0.6)^{10} + C(10,1)(0.4)^1(0.6)^9 \\ &\quad + C(10,2)(0.4)^2(0.6)^8 + C(10,3)(0.4)^3(0.6)^7 \\ &= .3823. \end{aligned} \tag{A.1.17}$$

Let us next examine the simple *stochastic process* of Poisson first described 150 years ago. This was the first microaxiomitization to introduce the important fourth dimension, time, into the probabilistic setting, although at the empirical level this had been done by John Graunt in 1662. Let us note, in advance, that each of the four axioms will generally be violated to some extent by most real-world applications. We wish to determine the probability that k occurrences of an event will occur between time 0 and time t. The axioms we shall use are

$$P(1 \text{ occurrence in } [t, t + \varepsilon]) = \lambda\varepsilon, \tag{A.1.18a}$$

$$P(\text{more than 1 occurrence in } [t, t + \varepsilon]) = o(\varepsilon) \tag{A.1.18b}$$

(where $\lim_{\varepsilon \to \infty} o(\varepsilon)/\varepsilon = 0$),

$$\begin{aligned} &P(k \text{ in } [t_1, t_2] \text{ and } m \text{ in } [t_3, t_4]) \\ &\quad = P(k \text{ in } [t_1, t_2]) P(m \text{ in } [t_3, t_4]) \quad \text{if } [t_1, t_2] \cap [t_3, t_4] = 0, \end{aligned} \tag{A.1.18c}$$

$$P(k \text{ in } [t_1, t_1 + s]) = P(k \text{ in } [t_2, t_2 + s]) \quad \text{for all } t_1, t_2, \text{ and } s. \tag{A.1.18d}$$

Then we may write

(A.1.19) $$\begin{aligned} P(k+1 \text{ in } [0, t+\varepsilon]) &= P(k+1 \text{ in } [0, t])P(0 \text{ in } [t, t+\varepsilon]) \\ &\quad + P(k \text{ in } [0, t])P(1 \text{ in } [t, t+\varepsilon]) + o(\varepsilon) \\ &= P(k+1, t)(1 - \lambda\varepsilon) + P(k, t)\lambda\varepsilon + o(\varepsilon) \end{aligned}$$

where $P(k, t) = P(k \text{ in } [0, t])$.

Then, we have

(A.1.20) $$\frac{P(k+1, t+\varepsilon) - P(k+1, t)}{\varepsilon} = \lambda[P(k, t) - P(k+1, t)] + \frac{o(\varepsilon)}{\varepsilon}.$$

Taking the limit as $\varepsilon \to \infty$, this gives Poisson's differential–difference equation:

(A.1.21) $$\frac{dP(k+1, t)}{dt} = \lambda[P(k, t) - P(k+1, t)].$$

Now taking $k = -1$, since we know that it is impossible for a negative number of events to occur, we have

(A.1.22) $$\frac{dP(0, t)}{dt} = -\lambda P(0, t).$$

So that

(A.1.23) $$P(0, t) = \exp(-\lambda t).$$

Substituting (A.1.23) in (A.1.21) for $k = 0$, we have

(A.1.24) $$\frac{dP(1, t)}{dt} = \lambda[\exp(-\lambda t) - P(1, t)],$$

with solution

(A.1.25) $$P(1, t) = \exp(-\lambda t)(\lambda t).$$

Continuing in this fashion for $k = 1$ and $k = 2$, we can conjecture the general formula of the Poisson distribution:

(A.1.26) $$P(k, t) = \frac{e^{-\lambda t}(\lambda t)^k}{k!}.$$

To verify that (A.1.26) satisfies (A.1.21), we simply substitute it into both sides of (A.1.21). Now let us note that the nonhappening of an event has special importance. $P(0, t)$ might describe, for example, the probability that an engine did not fail or that an epidemic did not occur. Let us suppose that we wish to generalize the constant failure rate λ to a $\lambda(t)$ which may vary with time; that is, we wish to relax the stationarity axiom (A.1.18d). Returning to (A.1.22), we would have

$$\frac{dP(0,t)}{dt} = -\lambda(t)P(0,t), \tag{A.1.27}$$

giving

$$P(0,t) = \exp\left(-\int_0^t \lambda(\tau)\,d\tau\right). \tag{A.1.28}$$

Now let us define the distribution function

$$F(t) = P(\text{happening on or before } t). \tag{A.1.29}$$

In the case of the nonstationary Poisson distribution, this becomes

$$F(t) = 1 - \exp\left(-\int_0^t \lambda(\tau)\,d\tau\right). \tag{A.1.30}$$

Defining the *density function* as

$$f(t) = \frac{dF(t)}{dt}, \tag{A.1.31}$$

we have for the example in (A.1.30)

$$f(t) = \exp\left(-\int_0^t \lambda(\tau)\,d\tau\right)\lambda(t). \tag{A.1.32}$$

In the special case when λ is constant, $f(t)$ is the density function of the *(negative) exponential distribution.*

Now in the case of the distribution function and the density function, it is not necessary that the argument be time. Any real value, such as income or tensile strength, will do. Let us suppose that we have a random variable X with distribution function $F(x)$, and let us further suppose that X is a *continu-*

ous random variable, that is, F is everywhere differentiable so that the density function $f(x)$ is well defined. We wish to find the distribution function of $Y = F(x)$. We shall achieve this by finding the distribution function of Y, say, $G(y)$.

$$G(y) = P(Y \leq y) = P(F(x) \leq y) = P(x \leq F^{-1}(y)) = y. \tag{A.1.33}$$

The next to the last statement in (A.1.33) is, in words, "the probability that X is less than or equal to that value than which X is less than or equal to y of the time." But this is simply y, and the equality follows. Equation (A.1.33) is the distribution function corresponding to the uniform distribution on the unit interval. The special tautology which gives rise to this distribution tells us that the distribution function of any continuous random variable is $U[0, 1]$. We observe that this fact makes the uniform distribution very important for simulation purposes. Let us suppose we have some means for generating random numbers from $U[0, 1]$. Then, if we have a random variable with distribution function $F(x)$, we can generate u from $U[0, 1]$ and then obtain a random X via

$$x = F^{-1}(u). \tag{A.1.34}$$

So far, we have examined some of the most important of the classical distributions: the hypergeometric, the binomial, the Poisson, the exponential, and the uniform. Each of these distributions has the right to claim the title of natural distribution, and because of results like that in (A.1.34), each can be deemed, in some sense, universal; that is, one could, if one insisted, do a great deal of modeling emphasizing exclusively just one of them. Because, for continuous random variables, the distribution function is strictly increasing, if two random variables have the same region where their density functions are positive, it is theoretically an easy matter to go from one distribution function to another. For example, let us suppose we have two continuous random variables X_1 and X_2 with distribution functions F_1 and F_2, respectively. Then, since both of these distribution functions, viewed as random variables themselves, are $U[0, 1]$, we may write

$$u = F_1(x_1) = F_2(x_2), \tag{A.1.35}$$

and hence

$$x_1 = F_1^{-1}(F_2(x_2)). \tag{A.1.36}$$

So, in some sense, every continuous distribution is universal. However, in practice, this "universality" is seldom used. For particular situations, it is generally the case that one particular distribution function appears the most natural.

We now consider another means by which a distribution may claim for itself universality. Suppose we have a number of random observations, say, $\{x_1, x_2, \ldots, x_n\}$, from the same distribution, having, say, distribution function $F(\cdot)$. A rather common operation to perform on such a set of random observations is to look at their average, the sample mean

$$\bar{x} = \frac{x_1 + x_2 + \cdots + x_n}{n}. \tag{A.1.37}$$

For almost any distribution one is likely to encounter, it turns out that for n large the sample mean has approximately a normal (also called DeMoivrean or Gaussian or Laplacian) density function; that is,

$$f(\bar{x}) = \frac{1}{\sqrt{2\pi\sigma_{\bar{x}}^2}} \exp\left(-\frac{1}{2\sigma_{\bar{x}}^2}(\bar{x} - \mu^2)\right), \quad \text{for } -\infty < \bar{x} < \infty. \tag{A.1.38}$$

In the above, the constant μ is simply the mean of the random variable X with distribution function $F(\cdot)$. The positive constant $\sigma_{\bar{x}}^2$ depends not only on the variance of X but also on the interrelationship between the various random sample values $\{x_1, x_2, \ldots, x_n\}$. Let us now examine the practical and useful notion of *independence.*

A.1.c. Dependence and Independence

Let us suppose we have two events A and B. Then if the joint probability of A and B occurring is given by

$$P(A \cap B) = P(A)P(B), \tag{A.1.39}$$

we say that A and B are *stochastically independent.*

Thus, if A is the event that the price of rye in the Chicago market will be higher tomorrow than it is today and B is the event that the number of meteors striking the earth 60 years from now will be greater than in the current year, we should expect that (A.1.39) will hold. Information to the effect that there will be a 60 year increase in annual meteor fall provides no help in predicting the next day increase in the price of rye. So the *joint probability* $P(A \cap B)$ is simply equal to the product of the *marginal probabilities* $P(A)$ and $P(B)$.

On the other hand, suppose that C is the event that the price of wheat is higher tomorrow in Chicago than it is today. Since there is a transfer of grain use between rye and wheat, if we know that the price of wheat will go up, then the probability that the price of rye will go up is greater than if we did not have the information about the increase in the price of wheat. Thus, we can introduce the notion of *conditional probability*, $P(A|C)$:

$$P(A \cap C) = P(C)P(A|C); \tag{A.1.40}$$

that is, the probability that A and C occur is equal to the probability that C occurs multiplied by the probability that A occurs given that C occurs. Furthermore, we can write (assuming $P(C)$ is not zero),

$$P(A|C) = \frac{P(A \cap C)}{P(C)}. \tag{A.1.41}$$

In the event that A and C are stochastically independent, the conditional probability $P(A|C)$ is simply equal to the marginal probability $P(A)$.

Now we should note that, depending on the circumstances of the actual problem, we should be able to write

$$P(A \cap C) = P(A)P(C|A). \tag{A.1.42}$$

Thus, the probability that the price of wheat and that of rye will rise tomorrow is equal to the probability that the price of wheat rises multiplied by the probability that the price of rye rises given that the price of wheat rises. Now the question of causation is a bit tricky. For example, rye is generally regarded as a less desirable grain than wheat. For the low end of many of the uses of wheat, rye can be substituted if wheat is in short supply and thus expensive. On the other hand, if the price of rye goes up, some of the rye for wheat substitution may be diminished. So one could argue that each price influences the other. In such a situation, A and C are dependent on each other, but it would probably be wrong to say that one was the cause, the other the effect. Indeed, it might be that both A and C were effects of an external cause, for example, a U.S. Weather Bureau forecast of long-term drought. The usual forms of conditional probability ((A.1.40) and (A.1.42)) make no natural distinction between cause and effect. Noting this rather bizarre formal symmetry between cause and effect, a nonconformist (Presbyterian) English clergyman, the Reverend Thomas Bayes, arrived at what appeared to be some rather startling consequences in the middle of the 18th century.

Let us suppose that, with reference to a particular problem, all possible "states of nature" can be written as the union of disjoint states:

$$\Omega = A_1 + A_2 + \cdots + A_n. \tag{A.1.43}$$

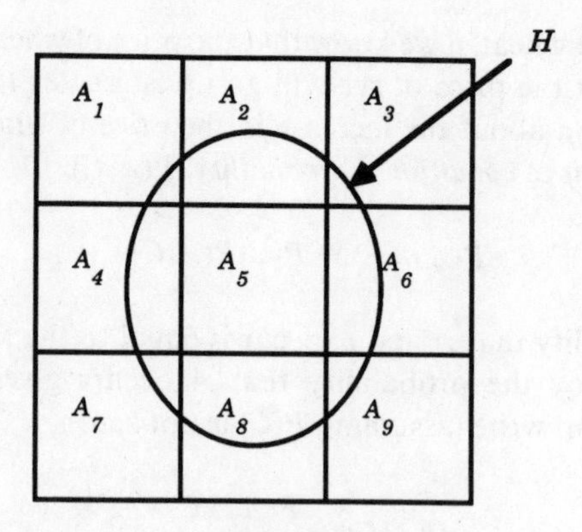

Figure A.1.1

Next, let us suppose that we have an additional piece of information, say, H, which has nonempty intersection with some of the $\{A_i\}$. Then Bayes wished to see how this information could be used to determine the probabilities of the various states of nature. A diagram of the situation is given in Figure A.1.1.

Let us try to find $P(A_1|H)$. We first note that we can write

(A.1.44) $$P(A_1|H) = \frac{P(A_1 \cap H)}{P(H)}.$$

But we also note that

(A.1.45) $$P(A_1 \cap H) = P(H|A_1)P(A_1).$$

Furthermore, since the $\{A_i\}$ are disjoint and exhaust all possible states of nature, we can write

(A.1.46) $$P(H) = P(A_1 \cap H) + P(A_2 \cap H) + \cdots + P(A_9 \cap H).$$

Adding to (A.1.46) the information from (A.1.45), we have

(A.1.47) $$P(H) = P(H|A_1)P(A_1) + P(H|A_2)P(A_2) + \cdots + P(H|A_9)P(A_9).$$

Then putting together (A.1.45) with (A.1.47) we have *Bayes' theorem*

(A.1.48) $$P(A_1|H) = \frac{P(H|A_1)P(A_1)}{P(H|A_1)P(A_1) + \cdots + P(H|A_9)P(A_9)}.$$

Let us now consider the rather remarkable implications of (A.1.48). We wish, on the basis of experimental information H, to infer the probability that the true state of nature is A_1. Equation (A.1.48) allows us to carry out this task, but only if we have knowledge of the *prior probabilities* $\{P(A_j)\}$. This strikes most people as bizarre, for it states that in order to make inferences from experimental data, we have to have information prior to that data. One common name given to information that is not data based is prejudice. The *posterior probability* $P(A_1|H)$ is the probability that the true state of nature is A_1 given the information H and the prior probabilities of each of the states. Once we compute the posterior probabilities of each of the states, these become our new prior probabilities.

To illustrate the practical use of Bayes' theorem, let us consider patients taking a test for a particular disease at a medical center. Historically, 5% of the patients tested for the disease have the disease. In 10% of the cases when the patient has the disease, the test falsely states that the patient does not have the disease. In 20% of the cases when the patient does not have the disease, the test falsely states that the patient has the disease. Suppose the patient tests positive for the disease. What is the posterior probability that the patient has the disease?

$$\text{(A.1.49)} \qquad P(D^+|T^+) = \frac{P(T^+|D^+)P(D^+)}{P(T^+|D^+)P(D^+) + P(T^+|D^-)P(D^-)}$$

$$= \frac{(.9)(.05)}{(.9)(.05) + (.2)(.95)} = .1915.$$

Not wishing unduly to alarm the patient, the medical center calls the patient in for another test. Again the test is positive. Then we go through the Bayes computation again using the recently computed posterior probabilities as the new prior probabilities. Thus, we have

$$\text{(A.1.50)} \qquad P(D^+|T^+) = \frac{P(T^+|D^+)P(D^+)}{P(T^+|D^+)P(D^+) + P(T^+|D^-)P(D^-)}$$

$$= \frac{(.9)(.1915)}{(.9)(.1915) + .(.2)(.8085)} = .5159.$$

This new posterior probability that the patient has the disease is sufficiently high that entry into the medical center for more intensive testing is probably indicated.

In many experimental situations there are no very good estimates for the prior probabilities. In such a case Bayes made the suggestion (*Bayes' axiom*) that we make the assumption that each of the states of nature are equally likely. Returning to (A.1.48), we note that the prior probabilities will simply cancel out

$$P(A_1|H) = \frac{P(H|A_1)P(A_1)}{P(H|A_1)P(A_1) + \cdots + P(H|A_9)\,\Pr(A_9)} = \frac{P(H|A_1)}{P(H|A_1) + \cdots + P(H|A_9)}. \tag{A.1.51}$$

Many will be willing to accept the logic of Bayes' theorem. However, an arbitrary selection of the prior probabilities, including those based on Bayes' axiom, are troubling. Note that in the testing example given earlier, if we had used the arbitrary prior probabilities of .50 for the disease and .50 for no disease, a positive test would have given a posterior probability of the disease of

$$P(D^+|T^+) = \frac{P(T^+|D^+)}{P(T^+|D^+) + P(T^+|D^-)} = \frac{.9}{.9 + .2} = .8182. \tag{A.1.52}$$

The value of .8182 is a far cry from the accurate value of .1915. However, it is intuitively clear that as we gather more and more data, as long as we did not start out with a zero probability for the true state of nature, we obtain posterior probabilities for the true state of nature which are better and better. In other words, even very bad "prejudices" will, sooner or later, be overcome by the addition of data. The eventual dominance of data over "prejudice" is one reason why the scientific method works.

Let us now note a semantic difficulty with the notion of stochastic independence. Let us suppose we have a population of voters, with an unknown proportion p of them being Republicans. A pollster samples 10 voters at random from the population and finds that each of them is a Republican. When he samples the 11th voter, is the probability that he will be a Republican stochastically independent of the fact that the first 10 voters sampled were Republicans? Let us suppose that the pollster, before sampling the 11th voter, had no information about the first 10 voters. Then the probability that the 11th voter will be a Republican would be simply p. What is the probability the voter will be a Republican given that 10 preceding voters were Republicans? The answer is still p. To see this, we simply follow (A.1.41):

(A.1.53) P (11th voter Republican | first 10 Republicans)

$$= P \text{ (all 11 voters Republicans)}/P \text{ (first 10 Republicans)}$$

$$= p^{11}/p^{10} = p.$$

That is, the fact that the first 10 voters turned out to be Republicans did not influence the basic mechanism of sampling.

More generally, let us suppose we have an (independent) random sample $\{X_1, X_2, \ldots, X_n\}$ of size n with probability function $p(x)$. We wish to find the joint density of $\{X_1, X_2, \ldots, X_n\}$. This is seen immediately to be

$$p(X_1, X_2, \ldots, X_n) = p(X_1)p(X_2)\cdots p(X_n). \tag{A.1.54}$$

We note the very important consequences of independence on expected values. Suppose $g_1, g_2, \ldots, g_n$ are functions of $X_1, X_2, \ldots, X_n$, respectively. Then

(A.1.55) $E_{1,2,\ldots,n}[g_1(X_1)g_2(X_2)\cdots g_n(X_n)]$

$$= \Sigma_{1,2,\ldots,n}[g_1(X_1)g_2(X_2)\cdots g_n(X_n)]p(X_1)p(X_2)\cdots p(X_n)$$

$$= \Sigma_1\, g_1(X_1)P(X_1)\sum_2 g_2(X_2)P(X_2)\cdots \sum_n g_n(X_n)P(X_n)$$

$$= E_1(g_1(X_1))E_2(g_2(X_2))\cdots E_n(g_n(X_n)).$$

Thus, stochastic independence of the $\{X_i\}$ implies that, for functions of the respective $\{X_i\}$, the expectation of the product is equal to the product of the expectations.

Let us now suppose we have two random variables X_1 and X_2, not necessarily independent, with expectations μ_1 and μ_2, respectively. Then we define the *covariance* of X_1 and X_2 as

$$\text{Cov}(X_1, X_2) = E[(X_1 - \mu_1)(X_2 - \mu_2)]. \tag{A.1.56}$$

Furthermore, we define the *correlation* between X_1 and X_2 as

$$\rho(X_1, X_2) = \frac{\text{Cov}(X_1, X_2)}{\sqrt{\sigma_1^2 \sigma_2^2}}. \tag{A.1.57}$$

Although covariances can take values from $-\infty$ to ∞, the correlation can take values only between -1 and 1. To see this, consider

$$0 \le E[a(X_1 - \mu_1) - (X_2 - \mu_2)]^2 = a^2\sigma_1^2 + \sigma_2^2 - 2a\,\text{Cov}(X_1, X_2), \tag{A.1.58}$$

where a is an arbitrary real constant which we select to be $\mathrm{Cov}(X_1, X_2)/\sigma_1^2$. Then we have, upon substitution, a variant of Cauchy's inequality:

$$\rho^2 \leq 1. \tag{A.1.59}$$

Next, if we consider the special case where X_1 and X_2 are stochastically independent, we note that

$$\begin{aligned} \mathrm{Cov}(X_1, X_2) &= E[(X_1 - \mu_1)(X_2 - \mu_2)] = E(X_1 - \mu_1)E(X - \mu_2) \\ &= 0. \end{aligned} \tag{A.1.60}$$

Then, from (A.1.57), we see immediately that the correlation between X_1 and X_2 is also zero in the case where the two random variables are stochastically independent.

A.1.d. Weak and Strong Laws of Large Numbers

Let us now consider a random i.i.d. (independent and identically distributed) sample of size n of a random variable X, $\{x_1, x_2, \ldots, x_n\}$. Let us suppose that the mean and variance of X are μ and σ^2, respectively. We wish to find the mean and expectation of the sample mean

$$\bar{x} = \frac{x_1 + x_2 + \cdots + x_n}{n}. \tag{A.1.61}$$

We note that

$$\begin{aligned} \mu_{\bar{x}} &= \frac{E(X_1 + X_2 + \cdots + X_n)}{n} = \frac{E(X_1) + E(X_2) + \cdots + E(X_n)}{n} \\ &= \frac{\mu + \mu + \cdots + \mu}{n} = \mu. \end{aligned} \tag{A.1.62}$$

In this derivation, we have not used independence or the fact that all the random variables have the same distribution, only the fact that they all have the same (finite) mean.

Next we shall derive the variance of $\bar{x}$

(A.1.63)

$$\begin{aligned} \sigma_{\bar{x}}^2 = \mathrm{Var}(\bar{x}) &= E[\bar{x} - \mu]^2 = E\left[\frac{(X_1 - \mu)}{n} + \frac{(X_2 - \mu)}{n} + \cdots + \frac{(X_n - \mu)}{n}\right]^2 \\ &= \sum_{j=1}^{n} \frac{E(X_j - \mu)^2}{n^2} + \text{terms like } E\left[\frac{(X_1 - \mu)(X_2 - \mu)}{n^2}\right]. \end{aligned}$$

Now, by independence, the expectation of the cross-product terms is zero. Thus, we have

$$\sigma_{\bar{x}}^2 = \frac{\sigma^2}{n}. \tag{A.1.64}$$

We note that in the derivation for the variance of the sample mean we have not used the fact that the $\{X_i\}$ are identically distributed, only the fact that they are independent and have the same μ and σ^2. The fact that the variability of $\bar{x}$ about its mean decreases as n increases is of key importance in experimental science. We shall develop this notion further below.

Let us consider a random variable Z and a nonnegative function $g(Z)$. Let

$$A = \{Z | g(Z) > \varepsilon^2\} \quad \text{and} \quad A^* = \{Z | g(Z) \leq \varepsilon^2\}. \tag{A.1.65}$$

Then,

$$E(g(Z)) = E_A(g(Z)) + E_{A^*}(g(Z)) \geq E_A(g(Z)) \geq \varepsilon^2 \Pr(A), \tag{A.1.66}$$

or

$$P(g(Z) > \varepsilon^2) \leq E(g(Z))/\varepsilon^2. \tag{A.1.67}$$

Let us consider the special case where $g(Z) = (Z - \mu_Z)^2$. Then, (A.1.67) becomes

$$P(|Z - \mu_Z| > \varepsilon) \leq \frac{\sigma_Z^2}{\varepsilon^2}. \tag{A.1.68}$$

Equation (A.1.68) is a form of *Chebyshev's inequality*. As a practical approximation device, it is not particularly useful. However, as an asymptotic device, it is invaluable. Let us consider the case where $Z = \bar{x}$. Then, (A.1.68) gives us

$$P(|\bar{x} - \mu| > \varepsilon) \leq \frac{\sigma^2}{n\varepsilon^2}, \tag{A.1.69}$$

or equivalently

$$P(|\bar{x} - \mu| \leq \varepsilon) > 1 - \frac{\sigma^2}{n\varepsilon^2}.$$

Equation (A.1.69) is a form of the *weak law of large numbers*. The WLLN tells us that if we are willing to take a sufficiently large sample, then we can obtain an arbitrarily large probability that $\bar{x}$ will be arbitrarily close to μ. The WLLN, as we have seen, is rather easy to prove constructively.

Although we shall omit the proof, a more powerful result, the *strong law of large numbers,* is available. By the SLLN, if the expectation of the absolute value of the random variable X is finite, then the sample mean must converge to $E(X)$ *almost surely*. That is, the sample mean converges in n to $E(X)$ except for cases that cannot happen. The rather practical advantage of the SLLN is that if H is some function, then

(A.1.70)
$$\lim_n H(\bar{x}) = H(\mu) \text{ a.s.},$$

where $\mu = E(X)$.

A.1.e. Moment-Generating Functions

Let us now consider the joint density function of a random i.i.d. (independent and identically distributed) sample $(x_1, x_2, \ldots, x_n)$ of a continuous random variable X having density function $f(\cdot)$. It is reasonable to define

(A.1.71)
$$\begin{aligned} f(x_1, x_2, \ldots, x_n) &= \lim_{\varepsilon_1, \varepsilon_2, \ldots \to 0} \frac{P[x_1 < X_1 \le x_1 + \varepsilon_1]}{\varepsilon_1} \frac{P[x_2 < X_2 \le x_2 + \varepsilon_2]}{\varepsilon_2} \cdots \\ &= \lim_{\varepsilon_1, \varepsilon_2, \ldots \to 0} \frac{f(x_1)\varepsilon_1}{\varepsilon_1} \frac{f(x_2)\varepsilon_2}{\varepsilon_2} \cdots \frac{f(x_n)\varepsilon_n}{\varepsilon_n} \\ &= f(x_1)f(x_2)\cdots f(x_n). \end{aligned}$$

The rather modest looking result in (A.1.71) was once mentioned to the author by the late Salomon Bochner as R. A. Fisher's greatest contribution to statistics. Note that it enables us to write the density of an n-dimensional random variable as the product of n one-dimensional densities. This attempt at dimensionality reduction is one of the recurring tasks of the applied mathematician.

In order to exploit the advantages of independence, we now consider the application of a simple version of the Fourier transformation. Suppose that X is a random variable with distribution function F_X. Then we define the *moment-generating function* $M_X(t)$ via

(A.1.72)
$$M_X(t) = E(e^{tX}).$$

Assuming that differentiation with respect to t commutes with the expectation operator, we have

$$M'_X(t) = E(Xe^{tX}) \tag{A.1.73}$$
$$M''_X(t) = E(X^2e^{tX})$$
$$\vdots$$
$$M^{(n)}_X(t) = E(X^ne^{tX}).$$

Setting t equal to zero, we see that

$$M^{(n)}_X(0) = E(X^n). \tag{A.1.74}$$

Thus, we see immediately the reason for the name moment-generating function (mgf). Once we have obtained $M_X(t)$, we can compute moments of arbitrary order (assuming they exist) by successively differentiating the mgf and setting the argument t equal to zero. As an example of this application, let us consider a random variable distributed according to the binomial distribution with parameters p and n. Then,

$$M_X(t) = \sum_0^n e^{tX}C(n, X)p^X(1-p)^{n-X} \tag{A.1.75}$$
$$= \sum_0^n C(n, X)(p^te)^X(1-p)^{n-X}.$$

Now recalling the binomial identity

$$\sum_0^n C(n, X)a^Xb^{n-X} = (a+b)^n, \tag{A.1.76}$$

we have $M'_X(t)$

$$M_X(t) = [pe^t + (1-p)]^n. \tag{A.1.77}$$

Next, differentiating with respect to t, we have

$$M'_X(t) = npe^t[pe^t + (1-p)]^{n-1}. \tag{A.1.78}$$

Then, setting t equal to zero, we have

$$E(X) = M'_X(0) = npe^0[pe^0 + (1-p)]^{n-1} = np. \tag{A.1.79}$$

Differentiating (A.1.78) again with respect to t, we have

(A.1.80)
$$M_X''(t) = npe^t[pe^t + (1-p)]^{n-1} + n(n-1)p^2e^{2t}[pe^t + (1-p)]^{n-2}.$$

Setting t equal to zero, we have

(A.1.81)
$$E(X^2) = M_X''(0) = np + n(n-1)p^2.$$

The possible mechanical advantages of the mgf are clear. One integration (summation) operation plus n differentiations will yield the first n moments of a random variable. The usual alternative would involve carrying out n integrations. Since it is usually easier to carry out symbolic differentiation than symbolic integration, the mgf can be a valuable tool for computing moments.

However, the moment-generating aspect of the mgf pales in importance to some of its properties relating to the summation of independent random variables. Let us suppose, for example, that we have n independently distributed random variables $X_1, X_2, \ldots, X_k$ with cdf's and mgf's $F_1, F_2, \ldots, F_k$ and $M_1, M_2, \ldots, M_k$, respectively. Suppose that we wish to investigate the distribution of the random variable

(A.1.82)
$$Z = c_1X_1 + c_2X_2 + \cdots + c_kX_k.$$

If we go through the complexities of transforming the variables and integrating all random variables except for Z, we can achieve the task. But consider using the moment-generating functions to achieve this task. Then we have

(A.1.83)
$$M_Z(t) = E[\exp\{t(c_1X_1 + c_2X_2 + \cdots + c_kX_k)\}].$$

Using the independence of the $\{X_j\}$ we may write

(A.1.84)
$$M_Z(t) = E[\exp\{tc_1X_1\}]E[\exp\{tc_2X_2\}]\cdots E[\exp\{tc_kX_k\}] = M_1(c_1t)M_2(c_2t)\cdots M_k(c_kt).$$

In many cases, we shall be able to use (A.1.84) to give ourselves immediately the distribution of Z. Suppose, for example, that we wish to pool the results of k pollsters who are estimating the proportion of voters favoring a particular candidate. If the true proportion is p and the sample sizes of each of the pollsters are $n_1, n_2, \ldots, n_k$, the numbers of individuals favoring the candidate in each of the subpolls are $X_1, X_2, \ldots, X_k$, respectively. The pooled number of voters favoring the candidate is simply

(A.1.85)
$$Z = X_1 + X_2 + \cdots + X_k.$$

Since each the ith pollsters is dealing with a binomial variable with parameters p and n_i, we have that the moment-generating function for Z is

$$\text{(A.1.86)}\quad \begin{aligned} M_Z(t) &= [pe^t + (1-p)]^{n_1}[pe^t + (1-p)]^{n_2}\cdots[pe^t + (1-p)]^{n_k} \\ &= [pe^t + (1-p)]^{n_1+n_2+\cdots+n_k}. \end{aligned}$$

We note that this is the moment-generating function of a binomial random variable with parameters p and $N = n_1 + n_2 + \ldots + n_k$. This result is just what we would have expected; namely, we simply pool the samples and treat the total number of individuals favoring the candidate as a binomial variate.

We note that in order to obtain the moment-generating function, we simply carry out an integration (or summation) where the integrand is the product of e^{tx} and the density function $f(\cdot)$. Thus, given the density function, we know what the mgf will be. It turns out that, under very general conditions, the same is true when going in the reverse direction; namely, if we know $M_X(t)$, we can compute a unique density that corresponds to it. The practical implication is that if we find a random variable with a mgf we recognize as corresponding to a particular density function, we know immediately that the random variable has the corresponding density function. It turns out that when summing independent random variables, (A.1.84) may imply that our life will be much easier if we work with the mgf. We shall show this shortly in the derivation of the *central limit theorem*.

Let us first consider the normal density function:

$$\text{(A.1.87)}\quad f(x) = \frac{1}{\sqrt{2\pi}\sigma}\exp\left(-\frac{1}{2\sigma^2}(x-\mu)^2\right), \quad \text{for} -\infty < x < \infty.$$

To satisfy ourselves that we have a true density function, we first note that $f(x) \geq 0$ for all x. We shall need to show that

$$\text{(A.1.88)}\quad \int_{-\infty}^{\infty} \frac{1}{\sqrt{2\pi}\sigma}\exp\left(-\frac{1}{2\sigma^2}(x-\mu)^2\right)dx = 1.$$

Initially, we make the transformation

$$\text{(A.1.89)}\quad z = \frac{x-\mu}{\sigma}.$$

Then the left side of (A.1.88) becomes

$$\text{(A.1.90)}\quad \int_{-\infty}^{\infty} \frac{1}{\sqrt{2\pi}}\exp\left(-\frac{z^2}{2}\right)dz = A.$$

We shall show that $A = 1$ by showing that $A^2 = 1$.

$$A^2 = \frac{1}{\sqrt{2\pi}} \int_{-\infty}^{\infty} \int_{-\infty}^{\infty} \exp[-\tfrac{1}{2}(z^2 + w^2)]\, dz\, dw. \tag{A.1.91}$$

Now transforming to polar coordinates, we have

$$r^2 = z^2 + w^2; \qquad \tan(\theta) = w/z.$$

Thus,

$$A^2 = \frac{1}{2\pi} \int_0^{\infty} \int_0^{2\pi} e^{-r^2/2} r\, dr\, d\theta = \frac{1}{2\pi} 2\pi \int_0^{\infty} e^{-r^2/2} r\, dr = 1. \tag{A.1.92}$$

We note that from our definition of a density function,

$$f(x) = \frac{dF(x)}{dx},$$

a density function must never be negative. Clearly, for the normal density function, this condition is met. Consequently, the two conditions for a function to be a proper probability density function are met for the normal density function.

Let us now find the mgf of a normal variate with mean μ and variance σ^2.

$$\begin{aligned} M_X(t) &= \frac{1}{\sqrt{2\pi\sigma^2}} \int_{-\infty}^{\infty} e^{tx} \exp\left(-\frac{1}{2\sigma^2}(x-\mu)^2\right) dx \\ &= \frac{1}{\sqrt{2\pi}\sigma} \int_{-\infty}^{\infty} \exp\left(-\frac{1}{2\sigma^2}(x^2 - 2\mu x - 2\sigma^2 t x + \mu^2)\right) dx \\ &= \frac{1}{\sqrt{2\pi}\sigma} \int_{-\infty}^{\infty} \exp\left(-\frac{1}{2\sigma^2}(x^2 - 2x(\mu + t\sigma^2) + \mu^2)\right) dx \\ &= \frac{1}{\sqrt{2\pi}\sigma} \int_{-\infty}^{\infty} \exp\left(-\frac{1}{2\sigma^2}(x - \mu^*)^2\right) dx \exp\left(t\mu + \frac{t^2\sigma^2}{2}\right), \end{aligned} \tag{A.1.93}$$

where $\mu^* = \mu + t\sigma^2$.

But recognizing that the integral is simply equal to unity, we see that the mgf of the normal distribution is given by

$$M_X(t) = \exp\left(t\mu + \frac{t^2\sigma^2}{2}\right). \tag{A.1.94}$$

A.1.f. Central Limit Theorem

We are now in a position to derive one version of the *central limit theorem.* Let us suppose we have a sample $\{x_1, x_2, \ldots, x_n\}$ of independently and identically distributed random variables with common mean μ and variance σ^2. We wish to determine, for n large, the approximate distribution of the sample mean

$$\bar{x} = \frac{1}{n}(x_1 + x_2 + \cdots + x_n).$$

We shall examine the distribution of the sample mean when put into "standard" form. Let

$$z = \frac{\bar{x} - \mu}{\sigma/\sqrt{n}} = \frac{x_1 - \mu}{\sqrt{n}\sigma} + \frac{x_2 - \mu}{\sqrt{n}\sigma} + \cdots + \frac{x_n - \mu}{\sqrt{n}\sigma}.$$

Now utilizing the independence of the $\{x_i\}$ and the fact that they are identically distributed with the same mean and variance, we can write

$$\text{(A.1.95)} \qquad \begin{aligned} M_z(t) = E(e^{tz}) &= \prod_{i=1}^{n} E\left[\exp\left(t\frac{x_i - \mu}{\sqrt{n}\sigma}\right)\right] \\ &= \left\{E\left[\exp\left(t\frac{x - \mu}{\sqrt{n}\sigma}\right)\right]\right\}^n \\ &= \left\{E\left[1 + t\frac{x - \mu}{\sqrt{n}\sigma} + \frac{t^2}{2}\frac{(x - \mu)^2}{n\sigma^2} + o\left(\frac{1}{n}\right)\right]\right\}^n \\ &= \left(1 + \frac{t^2}{2n}\right)^n \to e^{t^2/2} \quad \text{as } n \to \infty. \end{aligned}$$

But (A.1.95) is the mgf of a normal distribution with mean zero and variance one. Thus, we have been able to show that the sample mean of a (random) sample of size n of a random variable X, having mean μ and variance σ^2, becomes "close" to that of a normal distribution with mean μ and variance σ^2/n as n becomes large.

Let us consider, briefly, some of the enormous computational simplifications that are a consequence of the CLT. A pollster takes a (random) sample of size 400 from a population of voters and finds that 250 of them favor the Democratic candidate in an election. He wishes to use this information to make an inference about the proportion of voters who favor the candidate in the overall population. It is customary to make statements such as: "We can be 95% confident that p is between two values a and b." Now in this case we have observed a value of the sample mean equal to

$$\bar{x} = \hat{p} = \frac{250}{400} = .625. \tag{A.1.96}$$

But how shall we compute a and b? We could, for example, try to find a value $400b$ such that

$$\sum_{z=[400b]}^{z=400} C(400, z)(.625)^z(.375)^{400-z} = .025, \tag{A.1.97}$$

where $[400b]$ denotes the integer part of $400b$.
Similarly, we could estimate a by solving

$$\sum_{z=0}^{z=[400a]} C(400, z)(.625)^z(.375)^{400-z} = .025. \tag{A.1.98}$$

Using the central limit theorem, we have a much easier method of approach. The easiest way to remember the CLT for practical purposes is that if a statistic is the result of a summing process, then

$$Z = \frac{\text{statistic} - E(\text{statistic})}{\sqrt{\text{Var(statistic)}}}. \tag{A.1.99}$$

is approximately normally distributed with mean 0 and variance 1. The standard normal integral is not readily computable in closed form. However, using numerical approximations (equivalently, reading from tabulations of the normal distribution), we find that

$$\int_{-\infty}^{1.96} \frac{1}{\sqrt{2\pi}} e^{-z^2/2}\, dz = .975. \tag{A.1.100}$$

Quickly, then, in the present case, we obtain as our 95% *confidence interval for p*, $(.625 \pm 1.96\sqrt{[.625(.375)/400]})$, or $(.578, .672)$. That means, on the basis of the sample, we are 95% sure that at the time the sample was taken no fewer than 57.8% and no more than 67.2% of the electorate favored the Democratic candidate.

A.1.g. Conditional Density Functions

Let us now return to questions of interdependence between variables. If we are considering a two-dimensional random variable (X, Y), for example, it is quite natural to define the distribution function and density function in the continuous random variable case via

$$F(x, y) = \int_{-\infty}^{x} \int_{-\infty}^{y} f(x, y)\, dx\, dy. \tag{A.1.101}$$

Here

$$\frac{\partial^2 F(x, y)}{\partial x\, \partial y} = f(x, y). \tag{A.1.102}$$

Writing the statement of joint probability for small intervals in X and Y, we have

$$\begin{aligned} &\Pr(x \le X \le x + \varepsilon \cap y \le Y \le y + \delta) \\ &\quad = \Pr(x \le X \le x + \varepsilon) \Pr(y \le Y \le y + \delta | x \le X \le x + \varepsilon). \end{aligned} \tag{A.1.103}$$

Now exploiting the assumption of continuity of the density function, we can write

$$\begin{aligned} \int_{x}^{x+\varepsilon} \int_{y}^{y+\delta} f_{XY}(x, y)\, dx\, dy &= \int_{x}^{x+\varepsilon} f_X(x)\, dx \int_{y}^{y+\delta} f_{Y/x}(y)\, dy \\ &= \varepsilon\delta f_{XY}(x, y) = \varepsilon f_X(x) \delta f_{Y/x}(y). \end{aligned} \tag{A.1.104}$$

Here, we have used the terms f_{XY}, f_X, $f_{Y|x}$ to denote the *joint density function of X and Y*, the *marginal density function* of X, and the *conditional density function of Y given $X = x$*, respectively. This gives us immediately

$$f_{Y|x}(y) = \frac{f_{XY}(x, y)}{f_X(x)}. \tag{A.1.105}$$

To obtain the marginal density function f_X from the joint density function f_{XY} we integrate the joint density over all values of Y. Thus,

$$f_X(x) = \int_{-\infty}^{\infty} f_{XY}(x, y)\, dy. \tag{A.1.106}$$

A.1.h. Posterior Density Functions

Let us consider some consequences of Bayes' theorem (A.1.48). Typically, a density function $f(x)$ is characterized by one or more parameters $\Theta =$

$(\theta_1, \theta_2, \ldots, \theta_k)$. Thus, it is appropriate to write the density function as $f(x|\Theta)$. Observations $\{x_1, x_2, \ldots, x_n\}$ may be viewed as being dependent on the particular value of Θ. But, in a data analysis situation, it will be the $\{x_1, x_2, \ldots, x_n\}$, the observations, which are known. The characterizing parameter Θ will be the unknown. Therefore, it is tempting to use the Bayesian formulation to take the data and make an inference about Θ.

We may wish to reverse the order of conditioning and obtain the *posterior density function* $g(\Theta|x_1, x_2, \ldots, x_n)$. Using (A.1.106), we could write

$$g(\Theta|x_1, x_2, \ldots, x_n) = \frac{f(x_1, x_2, \ldots, x_n|\Theta)p(\Theta)}{h(x_1, x_2, \ldots, x_n)}. \tag{A.1.107}$$

In (A.1.107), $p(\Theta)$ is the *prior probability density function* of the unknown parameter Θ. The marginal density function $h(x_1, x_2, \ldots, x_n)$ is obtained by taking the joint density in the numerator and integrating out Θ. That is,

$$h(x_1, x_2, \ldots, x_n) = \int_{\theta_1}\int_{\theta_2}\cdots\int_{\theta_k} f(x_1, x_2, \ldots, x_n|\Theta)p(\theta_1, \theta_2, \ldots, \theta_k)\, d\theta_1\, d\theta_2 \cdots d\theta_k. \tag{A.1.108}$$

As we have noted earlier, there is something rather disturbing about the idea of requiring that we know a probability distribution on the unknown parameters before we take any data. However, as we shall see, for reasonable implementations of (A.1.107), as the sample size becomes large, the dependence of the posterior distribution in its regions of high density is largely on the data rather than on the prior distribution $p(\Theta)$.

A.1.i. Natural Priors

What is a "natural" prior distribution to take over the space of μ? One requirement we might make is that the prior be so chosen that the posterior distribution is of the same functional form as the prior. For example, if X is normally distributed with mean μ and variance σ^2 then it turns out that if we take as the prior distribution another normal distribution with mean μ_0 and variance σ_0^2, then (after straightforwardly and laboriously applying (A.1.107) and (A.1.108)) we find that the posterior density of μ is a normal density with mean and variance given by

$$\mu_n = \frac{1}{w_0 + w_n}(w_0\mu_0 + w_n\bar{x}); \qquad \sigma_n^2 = \frac{1}{w_0 + w_n} \tag{A.1.109}$$

where

$$w_0 = \frac{1}{\sigma_0^2} \quad \text{and} \quad w_n = \frac{n}{\sigma^2}.$$

There is much that is satisfying about (A.1.109). We note that for all sizes of the sample, from no data at all to a data set of arbitrarily large size, the posterior distribution of μ is normal. We observe that for small sample sizes, we place rather more reliance on the mean of the prior distribution μ_0. However, as the sample size becomes large, we are placing almost all our reliance on the sample mean, little on the posterior mean. We also note that the variance of the posterior distribution is always decreasing as the sample size increases. For no data at all, it is precisely the variance of the prior distribution σ_0^2. As the sample size becomes very large, however, the posterior variance becomes essentially the variance of the sample mean, namely, σ^2/n. Moreover, we note that the variance is going to zero as n goes to infinity. That is, the posterior distribution is approaching a Dirac "spike" at the sample mean

$$g(\mu|\bar{x}) \underset{n}{\rightarrow} \delta(\mu - \bar{x}) \tag{A.1.110}$$

where

$$\delta(\mu - \bar{x}) = 0 \quad \text{if } \mu \neq \bar{x} \quad \text{and} \quad \int_{-\infty}^{\infty} \delta(\mu - \bar{x})\, d\mu = 1. \tag{A.1.111}$$

Furthermore, we note that, as a consequence of the central limit theorem, we might be tempted to use the Bayesian argument above more generally. That is, whenever we wish to estimate the mean of a distribution of known variance, we might, for samples of moderate to large size, be content to utilize the above argument, even if the underlying distribution is not normal.

A.1.j. Posterior Means

We observe that the Bayesian argument is strongly oriented toward a psychological point of view which many will find unnatural. Theoretically, the Bayesian does not say, "on the basis of prior information and a data set, we feel that the mean is approximately so-and-so." Rather, the Bayesian supposedly says, "on the basis of prior information and a data set, we feel that the mean has the posterior density indicated." As a practical matter, in order to communicate with the rest of us, the Bayesian is frequently willing to pick one value of μ as approximately the true value of the mean. The choices that might be made for this value are numerous. For example, we might simply choose μ_n, the posterior mean. Such a choice has much to recommend it. It can be shown, under very general conditions, that the posterior mean of a parameter θ minimizes the expected squared deviation of an estimator from the true value. To see this let us consider the estimation of an unknown parameter θ characterizing a density function $f(X|\theta)$ on the basis of a sample $\{x_1, x_2, \ldots, x_n\}$ and a prior density $p(\theta)$ on the parameter. We shall seek the estimate θ_B a function of the data and the parameters characterizing the prior density, which minimizes the expected squared deviation of θ_B from the actual θ integrated over the θ space.

$$\text{(A.1.112)} \quad \int_{x_1}\int_{x_2}\cdots\int_{x_n}\int_{\theta}(\theta_B-\theta)^2 f(x_1, x_2, \ldots, x_n|\theta)p(\theta)\,d\theta\,dx_1\,dx_2\cdots dx_n.$$

Now recalling that θ_B is to be a function of the sample only, we note that we need only concern ourselves, for each value of $\{x_1, x_2, \ldots, x_n\}$, with the value of θ_B which minimizes the inner integral over θ space. Differentiating this inner integrand with respect to θ_B and setting the derivative equal to zero, we have

$$\text{(A.1.113)} \quad \int_{\theta}(\theta_B-\theta)f(x_1, x_2, \ldots, x_n|\theta)p(\theta)\,d\theta = 0.$$

This gives immediately as our estimate for θ the posterior mean

$$\text{(A.1.114)} \quad \theta_B = \frac{\int_{\theta}\theta f(x_1, x_2, \ldots, x_n|\theta)p(\theta)\,d\theta}{\int_{\theta} f(x_1, x_2, \ldots, x_n|\theta)p(\theta)\,d\theta} = \int_{\theta}\theta g(\theta|x_1, x_2, \ldots, x_n)\,d\theta.$$

The posterior mean θ_B is usually referred to as the *Bayes estimate* for θ.

A.1.k. Maximum Likelihood

Another estimate that we might have selected is the *posterior mode,* that is, the value of θ for which the posterior density $g(\theta|x_1, x_2, \ldots, x_n)$ is a maximum. It should be noted that the controversial notion of assuming prior knowledge about the unknown parameter θ before the data are taken is difficult to avoid. Suppose that one wishes to take the position of making no prior assumption concerning the actual value of θ. The Bayesian framework allows this kind of approach provided that one is willing to specify with some precision what prior ignorance means.

Let us consider one such description of prior ignorance. Let us suppose that the prior density is given by

$$p(\theta) = \frac{1}{b-a} \quad \text{if } a \le \theta \le b \tag{A.1.115}$$
$$= 0 \qquad \text{otherwise.}$$

Here we assume that a is so small and b is so large that it is impossible that θ could be outside the interval $[a, b]$. This would be one possible interpretation of prior ignorance. Then we would have as the posterior density of θ

$$g(\theta|x_1, x_2, \ldots, x_n) = \frac{f(x_1, x_2, \ldots, x_n|\theta)/(b-a)}{= \int_a^b [f(x_1, x_2, \ldots|\theta^*)/(b-a)]\, d\theta^*} \tag{A.1.116}$$
$$= K(x_1, x_2, \ldots, x_n) f_n(x_1, x_2, \ldots, x_n|\theta).$$

In order to obtain the posterior mode of the posterior distribution with this "noninformative prior," we simply maximize the right-hand side of (A.1.116) with respect to θ. But we note that the only term on the right-hand side which involves θ is the *likelihood* $f_n(x_1, x_2, \ldots, x_n|\theta)$. In the important case of independent sampling, we have

$$f_n(x_1, x_2, \ldots, x_n|\theta) = \prod_{j=1}^{n} f(x_j|\theta). \tag{A.1.117}$$

Thus, using one definition of prior ignorance, we see that the posterior mode can be found by looking for the *maximum likelihood estimator*

$$\hat{\theta} = \theta \quad \text{such that } \prod_{j=1}^{n} f(x_j|\theta) \text{ is maximized.} \tag{A.1.118}$$

Using the notion that it is the data that are fixed, not θ, from the standpoint of the observer, it is common to write $L(\theta|x_1, x_2, \ldots, x_n)$ instead of $f_n(x_1, x_2, \ldots, x_n|\theta)$ for the likelihood. A customary technique for finding the maximum likelihood estimator of θ is to differentiate $f_n(x_1, x_2, \ldots, x_n|\theta)$ with respct to θ and set the derivative to zero. We shall follow this convention below, while noting that the assumption of uniqueness for the estimator so obtained is frequently violated in practice, particularly in the case where θ is multidimensional. The maximum likelihood estimator so obtained will be denoted by $\hat{\theta}_n$.

The interest in maximum likelihood estimation flows, in large measure, from its desirable properties in the large sample case. We shall demonstrate some of these below. Let us consider the case where the probability density function $f(\cdot|\theta)$ is regular with respect to its first θ derivative in the parameter space Θ; that is,

$$E\left[\frac{\partial \log f(x|\theta)}{\partial \theta}\right] = \frac{\partial}{\partial \theta}\int_{-\infty}^{\infty} f(x|\theta)\,dx = 0. \tag{A.1.119}$$

Let us assume that for any given sample $(x_1, x_2, \ldots, x_n)$, Eq. (A.1.120) below has a unique solution.

$$\left[\frac{\partial}{\partial \theta}\log f_n(x_1, x_2, \ldots, x_n)\right]_{\hat{\theta}_n} = \sum_{j=1}^{n}\left[\frac{\partial}{\partial \theta}\log f(x_j|\theta)\right]_{\hat{\theta}_n} = 0. \tag{A.1.120}$$

Let us suppose our sample $(x_1, x_2, \ldots, x_n)$ has come from $f(\cdot|\theta_0)$; that is, the actual value of θ *is the point* $\theta_0 \in \Theta$. We wish to examine the stochastic convergence of $\hat{\theta}_n$ to θ_0. We shall assume that

$$\begin{aligned} B^2(\theta, \theta) &= \operatorname{Var}\left[\frac{\partial \log f(x|\theta)}{\partial \theta}\right] \\ &= \int_{-\infty}^{\infty}\left(\frac{\partial \log f(x|\theta)}{\partial \theta}\right)^2 f(x|\theta)\,dx < \infty. \end{aligned} \tag{A.1.121}$$

We now examine

$$H(\theta_0, \theta) = \int_{-\infty}^{\infty} [\log f(x|\theta)] f(x|\theta_0)\,dx. \tag{A.1.122}$$

Considering the second difference of $H(\theta_0)$ about θ_0, we have, using the strict concavity of the logarithm and Jensen's inequality (taking care that $[\theta_0 - h, \theta_0 + h] \in \Theta$),

(A.1.123)

$$\begin{aligned} &\Delta^2_{\theta_0,h} H(\theta_0, \theta) \\ &\quad = \int \log[f(x|\theta_0 + h)] f(x|\theta_0)\,dx \\ &\qquad + \int \log[f(x|\theta_0 - h)] f(x|\theta_0)\,dx - 2\int \log[f(x|\theta_0)] f(x|\theta_0)\,dx \end{aligned}$$

$$= \int \log\left[\frac{f(x|\theta_0 + h)}{f(x|\theta_0)}\right] f(x|\theta_0)\,dx + \int \log\left[\frac{f(x|\theta_0 - h)}{f(x|\theta_0)}\right] f(x|\theta_0)\,dx$$

$$< \log \int \frac{f(x|\theta_0 + h)}{f(x|\theta_0)} f(x|\theta_0)\,dx + \log \int \frac{f(x|\theta_0 - h)}{f(x|\theta_0)} f(x|\theta_0)\,dx$$

$$= \log 1 + \log 1 = 0.$$

Thus, we have shown that

$$\Delta^2_{\theta_0,h} H(\theta_0, \theta) < 0. \tag{A.1.124}$$

But then,

$$A(\theta_0, \theta) = \int \left[\frac{\partial \log f(x|\theta)}{\partial \theta}\right] f(x|\theta_0)\,dx \tag{A.1.125}$$

is strictly decreasing in a neighborhood $(\theta_0 - \delta, \theta + \delta) \in \Theta$ about θ_0.

We note that

$$\frac{1}{n}\frac{\partial}{\partial\theta} \log f_n(x_1, x_2, \ldots, x_n|\theta) = \frac{1}{n}\sum \frac{\partial}{\partial\theta} \log f(x_j|\theta) \tag{A.1.126}$$

is the sample mean of a sample of size n. Thus, by the strong law of large numbers (see (A.1.70)), we know that, almost surely,

$$\frac{1}{n}\frac{\partial}{\partial\theta} \log f_n(x_1, x_2, \ldots, x_n) \to \int \frac{\partial}{\partial\theta}[\log f(x|\theta)] f(x|\theta_0)\,dx = A(\theta_0, \theta). \tag{A.1.127}$$

Moreover, by (A.1.119), we have that $A(\theta_0, \theta_0) = 0$. Thus, for any $\varepsilon > 0$, $0 < \delta' < \delta$, there may be found an $n(\delta', \varepsilon)$, such that the probability exceeds $1 - \varepsilon$ that the following inequalities hold for all $n > n(\delta', \varepsilon)$:

$$\frac{1}{n}\frac{\partial}{\partial\theta} \log f_n(x_1, x_2, \ldots, x_n) > 0 \quad \text{if } \theta = \theta_0 - \delta'; \tag{A.1.128}$$

$$\frac{1}{n}\frac{\partial}{\partial\theta} \log f_n(x_1, x_2, \ldots, x_n) < 0 \quad \text{if } \theta = \theta_0 + \delta'.$$

Consequently, assuming $(\partial/\partial\theta) \log f_n(x_1, x_2, \ldots, x_n)$ is continuous in θ over $(\theta_0 - \delta', \theta + \delta')$, we have, for some θ in $(\theta_0 \pm \delta')$ for all $n > n(\delta', \varepsilon)$,

$$P\left(\frac{1}{n}\frac{\partial}{\partial\theta} \log f_n(x_1, x_2, \ldots, x_n|\theta) = 0|\theta_0\right) > 1 - \varepsilon. \tag{A.1.129}$$

Recall that we have assumed that $\hat{\theta}_n$ uniquely solves (A.1.120). Thus, we have shown that the maximum likelihood estimator converges almost surely to θ_0.

Now since f_n is a density, we have

$$\int_{-\infty}^{\infty} \cdots \int_{-\infty}^{\infty} f_n(x_1, x_2, \ldots, x_n|\theta) \prod_{j=1}^{n} dx_j = 1. \tag{A.1.130}$$

Assuming (A.1.130) is twice differentiable under the integral sign, we have upon one differentiation

$$\begin{aligned} \int \cdots \int \frac{\partial f_n}{\partial \theta} \prod dx_j &= \int \cdots \int \left(\frac{1}{f_n} \frac{\partial f_n}{\partial \theta}\right) f_n \prod dx_j \\ &= \int \cdots \int \frac{\partial \log f_n}{\partial \theta} f_n \prod dx_j = 0. \end{aligned} \tag{A.1.131}$$

Differentiating again, we have

$$\begin{aligned} &\int \cdots \int \left[f_n \frac{\partial}{\partial \theta}\left(\frac{1}{f_n} \frac{\partial f_n}{\partial \theta}\right) + \left(\frac{\partial f}{\partial \theta}\right)^2 \frac{1}{f_n} \right] \prod dx_j \\ &= \int \cdots \int \left[\frac{\partial^2 \log f_n}{\partial \theta^2} + \left(\frac{1}{f_n} \frac{\partial f_n}{\partial \theta}\right)^2 \right] f_n \prod dx_j = 0. \end{aligned} \tag{A.1.132}$$

This yields

$$E\left[\frac{\partial^2 \log f_n}{\partial \theta^2}\right] = -E\left[\left(\frac{\partial \log f_n}{\partial \theta}\right)^2\right]. \tag{A.1.133}$$

Now if we have an unbiased estimator for θ, say θ^*, then

$$\int \cdots \int (\theta^*(x_n) - \theta) f_n(x_1, x_2, \ldots, x_n|\theta) \prod dx_j = 0. \tag{A.1.134}$$

Differentiating with respect to θ, we have

$$\int \cdots \int (\theta^* - \theta) \frac{\partial \log f_n}{\partial \theta} f_n \prod dx_j = 1, \tag{A.1.135}$$

or

$$E\left[(\theta^* - \theta)\left(\frac{\partial \log f_n}{\partial \theta}\right)\right] = 1. \tag{A.1.136}$$

This gives us, by the Cauchy inequality (A.1.59),

$$E[(\theta^* - \theta)^2] \geq \frac{1}{E[(\partial \log f_n/\partial\theta)^2]} = \frac{-1}{E[\partial^2 \log f_n/\partial\theta^2]}. \tag{A.1.137}$$

The result in (A.1.137) gives us a lower bound for the mean square error of an unbiased estimator. Since the observations are independent and identically distributed, we can easily show that

$$E[(\theta^* - \theta)^2] \geq \frac{-1}{nE[(\partial^2 \log f(x|\theta)/\partial\theta^2]}. \tag{A.1.138}$$

Next, let us assume that the first two derivatives with respect to θ of $f_n(x_1, x_2, \ldots, x_n|\theta)$ exist in an interval about the true value θ_0. Furthermore, let

$$E\left[\frac{\partial \log f_n}{\partial\theta}\right] = 0, \tag{A.1.139}$$

and let

$$B_n^2(\theta) = E\left[\left(\frac{\partial \log f_n}{\partial\theta}\right)^2\right] = -E\left[\frac{\partial^2 \log f_n}{\partial\theta^2}\right] = -nE\left[\frac{\partial^2 \log f}{\partial\theta^2}\right]$$

be nonzero for all θ in the interval. We shall show that, asymptotically, the maximum likelihood estimator for θ is asymptotically Gaussian with mean θ_0 and variance equal to the Cramer–Rao lower bound $1/B_n^2(\theta_0)$.

Taking a Taylor's series expansion about θ_0, we have

$$\left(\frac{\partial \log f_n}{\partial\theta}\right)_{\hat{\theta}_n} = \left(\frac{\partial \log f_n}{\partial\theta}\right)_{\theta_0} + (\hat{\theta}_n - \theta_0)\left(\frac{\partial^2 \log f_n}{\partial\theta^2}\right)_{\theta_n'} \tag{A.1.140}$$

where θ_n' is between θ_0 and $\hat{\theta}_n$.

Rewriting (A.1.140), we have

$$(\hat{\theta}_n - \theta_0)B_n(\theta_0) = \frac{\left(\dfrac{\partial \log f_n}{\partial\theta}\right)_{\theta_0}\Big/ B_n(\theta_0)}{\left(\dfrac{\partial^2 \log f_n}{\partial\theta^2}\right)_{\theta_n'}\Big/(-B_n^2(\theta_0))} \tag{A.1.141}$$

$$= \frac{\sum_{j=1}^{n}\left(\dfrac{\partial \log f(x_j)}{\partial\theta}\right)_{\theta_0}\Big/ B_n(\theta_0)}{\dfrac{1}{n}\sum_{j=1}^{n}\left(\dfrac{\partial^2 \log f(x_j)}{\partial\theta^2}\right)_{\theta_n'}\Big/ E\left[\left(\dfrac{\partial^2 \log f}{\partial\theta^2}\right)_{\theta_0}\right]}.$$

But we have shown in (A.1.129) that $\hat{\theta}_n$ converges almost surely to θ_0. Hence, in the limit, the denominator on the right-hand side of (A.1.141) becomes unity. Moreover, we note that

$$\frac{\partial \log f_n}{\partial \theta} = \sum_{j=1}^{n} \frac{\partial \log f(x_j)}{\partial \theta} = \sum_{j=1}^{n} y_j, \tag{A.1.142}$$

where the $\{y_j\}$ are independent random variables with mean zero and variance

$$E\left[\left(\frac{\partial \log f(x|\theta)}{\partial \theta}\right)^2_{\theta_0}\right] = \frac{1}{n} B_n^2(\theta_0). \tag{A.1.143}$$

Thus, we have the rather amazing result that the maximum likelihood estimator $\hat{\theta}_n$ converges almost surely to θ_0 and that its asymptotic distribution is Gaussian with a variance equal to the Cramer–Rao lower bound for the variance of an unbiased estimator for θ_0. Little wonder that maximum likelihood estimation is so widely used *when it can be*. The reality is that, for most problems, it is so difficult to obtain the likelihood explicitly, that maximum likelihood cannot be used.

A.1.l. Simulation-Based Minimum χ^2 Estimation

For those situations where the likelihood is unknown, it may be appropriate to turn to an older, seldom employed technique developed by Karl Pearson. An application of this approach is discussed extensively in Section 3.3.

Let us suppose that we have a stochastic mechanism $S(X, \Theta)$, where X and Θ may be multidimensional. The parameter Θ specifies the process. The X are the data generated by the process. For purposes of explication, let us consider the case where X is one-dimensional. Let us suppose we have a set of 32 data points as shown in Figure A.1.2. We know the data but not the parameter Θ which describes the process. We wish to use the data to estimate the unknown parameter. Pearson suggested that one way to estimate Θ might be first to bin the data and then see if a particular value of Θ gave the same estimate for the proportion which should fall in each bin as that actually observed. Let us suppose that we bin the above data such that we have four observations in each bin (Figure A.1.3). Division points will be equidistant between boundary data points. The proportion of points p_i in each bin is 0.125.

Now, making a guess for Θ, we generate, say, 16 pseudo-observations as indicated by the solid circles as shown in Figure A.1.4. This gives the simulated proportions $\pi_1 = \pi_2 = \frac{1}{16}$; $\pi_4 = \pi_5 = \pi_7 = \pi_8 = \frac{2}{16}$; $\pi_3 = \pi_6 = \frac{3}{16}$. The criterion function proposed by Pearson is

$$\chi^2(\Theta) = \sum_{i=1}^{8} \frac{(p_i - \pi_i)^2}{\pi_i}. \tag{A.1.144}$$

Figure A.1.2

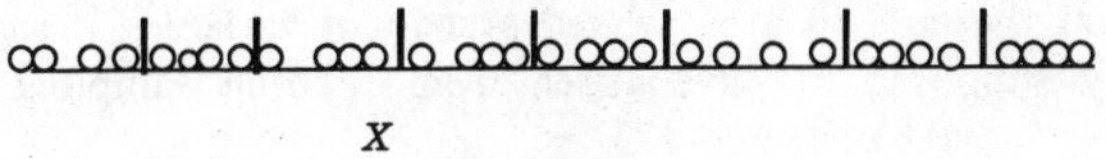

Figure A.1.3

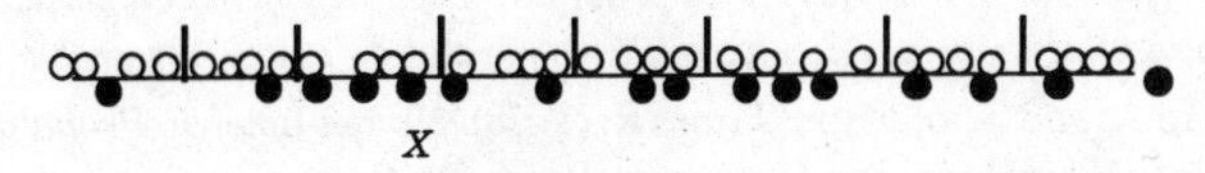

Figure A.1.4

In the case above, we have

$$\text{(A.1.145)} \quad \chi^2(\Theta) = 2\frac{(\frac{1}{8} - \frac{1}{16})^2}{\frac{1}{16}} + 4\frac{(\frac{1}{8} - \frac{1}{8})^2}{\frac{1}{8}} + 2\frac{(\frac{1}{8} - \frac{3}{16})^2}{\frac{3}{16}} = \frac{1}{6}.$$

In the original formulation by Pearson, nearly a century ago, simulation did not enter the picture. Computers were not even fantasized. The procedure for obtaining the π_j was via

$$\text{(A.1.146)} \quad \pi_j = \int_{j\text{th bin}} f(X|\Theta)\,dX.$$

In other words, the Pearsonian minimum chi-square formulation, like the subsequent maximum likelihood procedure of Fisher, of computational necessity, assumed knowledge of the underlying probability density function. In Section 3.3, we demonstrate how both notions can be generalized to the more realistic situation where we do not know the density function. But the procedure is more in the spirit of Pearson than that of Fisher, since we follow Pearson's intuitive notion of picking the parameter values which produce observations that best mimic the data.

References

Bayes, Thomas (1764/1963). *Facsimiles of Two Papers by Bayes*, Hafner, New York.

Fisher, Ronald A. (1922). On the mathematical foundations of theoretical statistics,

Philosophical Transactions of the Royal Society of London, Series A, **222**, pp. 309–368.

Kendall, Maurice G. and Stuart, Alan (1961). *The Advanced Theory of Statistics*, Vol. **2**, Hafner, New York.

Pearson, Karl (1900). On the criterion that a given system of deviations from the probable in the case of a correlated system of variables is such that it can reasonably supposed to have arisen from random sampling, *Philosophical Magazine* (5th Series), **50**, 157–175.

Pfeiffer, Paul E. and Schum, David A. (1973). *Introduction to Applied Probability*, Academic Press, New York.

Poisson, S. D. (1837). *Recherches sur la probabilité des jugements en matière criminelle et en matière civile, precédées des réglés générales du calcul des probabilités*, Paris.

Tapia, Richard A. and Thompson, James R. (1978). *Nonparametric Probability Density Estimation*, Johns Hopkins University Press, Baltimore.

Wilks, Samuel S. (1962). *Mathematical Statistics*, Wiley, New York.

A.2. A SIMPLE OPTIMIZATION ALGORITHM THAT USUALLY WORKS

There exist literally hundreds of optimization algorithms. Unfortunately, most of these are not of much use to the applied worker, who is confronted with data contaminated by noise. Below, we consider the polytope algorithm of Nelder and Mead (1965). Let us suppose that we are confronted with the task of minimizing a function F, which depends on k variables—say $(X_1, X_2, \ldots, X_k)$. We wish to find the value that approximately minimizes F. To accomplish this task we start out with $k + 1$ trial values of $(X_1, X_2, \ldots, X_k)$, of rank k. We evaluate F at each of these values and identify the three values where F is the minimum (B), the maximum (W), and the second largest ($2W$). We average all the X points except W to give us the centrum C. We show a pictorial representation for the case where $k = 2$. Note, however, that the flowchart below works for any $k \geq 2$. This is due to the fact that our moves are oriented around the best, the worst, and the second worst points. Note that the algorithm first moves by a succession of expansion steps to a region where F is small. Once this has been achieved, the procedure essentially works by the simplex of vertices surrounding and collapsing in on the optimum (slow in high dimensions). Clearly, the procedure does not make the great leaps characteristic of Newton procedures. It is slow, particularly if k is large. Even during the expansion steps, the algorithm moves in a zig zag, rather than in straight thrusts to the target. But the zig zag and envelopment strategies help avoid charging off on false trails. This algorithm has been presented because of its robustness to noise in the objective function (of particular importance when dealing with such approaches as that delineated in Section 3.3), the ease with which it can be programmed for the microprocessor, and the very practical fact that once learned, it is easy to remember.

Expansion

```
P = C + γR(C − W) (where typically γR = γE = 1)
If F(P) < F(B), then
        PP = P + γE(C − W)                      [a]
        If F(PP) < F(P), then
          Replace W with PP as new vertex        [c]
        Else
          Accept P as new vertex                 [b]
        End If
Else
  If F(P) < F(2W), then Accept P as new vertex   [b]
Else
```

Contraction

```
If F(W) < F(P) then
        PP = C + γC(W − B) (typically γC = 1/2)         [a*]
        If F(PP) < F(W) then
          Replace W with PP as new vertex              [b*]
        Else
          Replace W with (W + B)/2 and 2W with
          (2W + B)/2 (total contraction)               [c*]
        End If
Else
```

Contraction

```
If F(2W) < F(P), then
        PP = C + γC(P − B)                             [aa]
        If F(PP) < F(P), then
          Replace W with PP as new vertex              [bb]
        Else
          Replace W with (W + B)/2 and 2W with
          (2W + B)/2 (total contraction)               [cc]
Else
        Replace W with P
        End If
End If
```

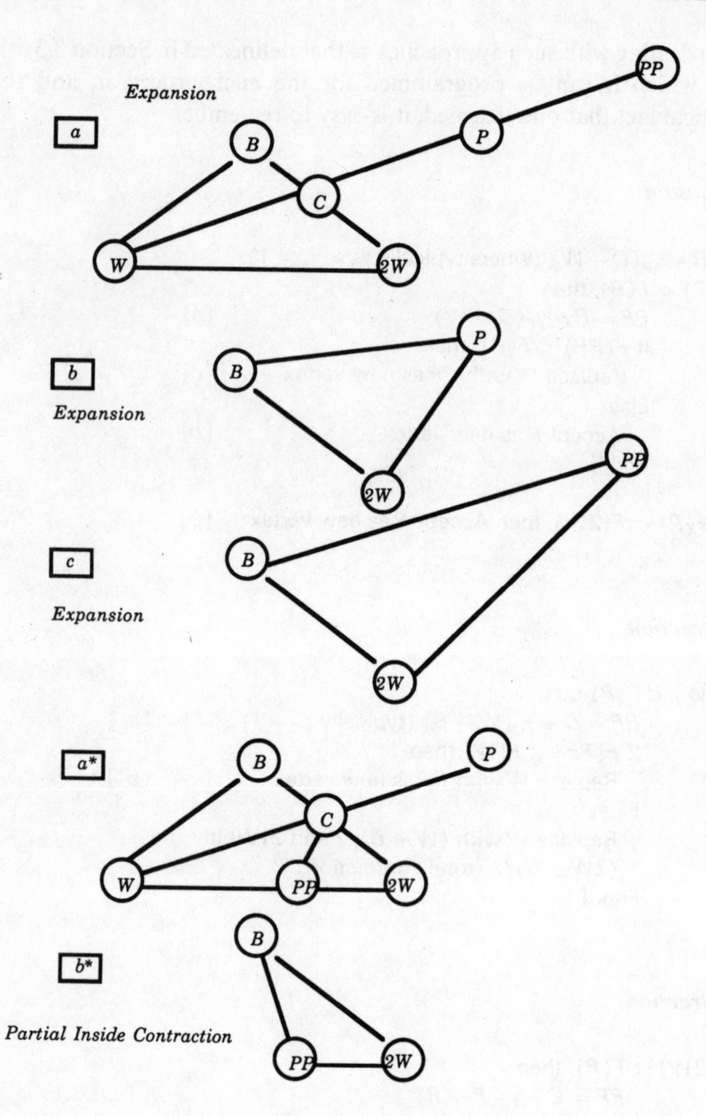
Expansion
a
B
P
PP
C
W
2W
b
Expansion
P
B
2W
PP
c
Expansion
B
2W
PP
a*
B
P
C
W
PP
2W
b*
Partial Inside Contraction
B
PP
2W

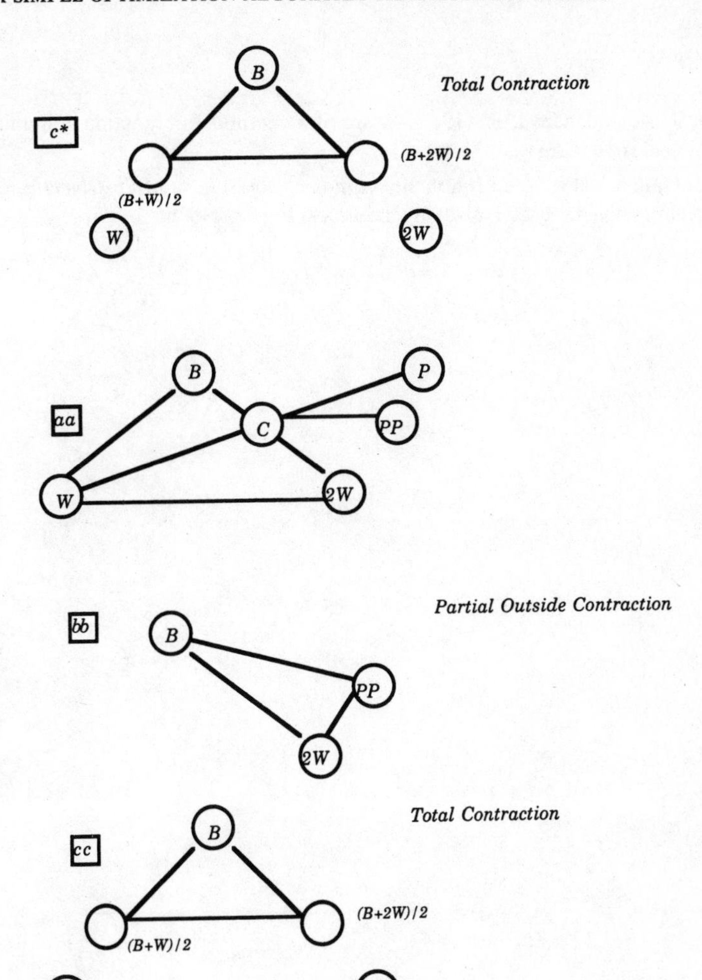

c*
B
Total Contraction
(B+2W)/2
(B+W)/2
W
2W
aa
B
P
C
PP
W
2W
Partial Outside Contraction
bb
B
PP
2W
Total Contraction
cc
B
(B+2W)/2
(B+W)/2
W
2W

References

Nelder, J. A. and Mead, R. (1965). A simplex method for function minimization. *Computational Journal,* **7,** 308–313.

Woods, Daniel J. (1985). *An Interactive Approach for Solving Multi-Objective Optimization Problems,* Rice University Technical Report No. 85-5.

Index

Applied Probability and Statistics (Continued)

HAND • Discrimination and Classification
HEIBERGER • Computation for the Analysis of Designed Experiments
HOAGLIN, MOSTELLER and TUKEY • Exploring Data Tables, Trends and Shapes
HOAGLIN, MOSTELLER, and TUKEY • Understanding Robust and Exploratory Data Analysis
HOCHBERG and TAMHANE • Multiple Comparison Procedures
HOEL • Elementary Statistics, *Fourth Edition*
HOEL and JESSEN • Basic Statistics for Business and Economics, *Third Edition*
HOGG and KLUGMAN • Loss Distributions
HOLLANDER and WOLFE • Nonparametric Statistical Methods
IMAN and CONOVER • Modern Business Statistics
JESSEN • Statistical Survey Techniques
JOHNSON • Multivariate Statistical Simulation
JOHNSON and KOTZ • Distributions in Statistics
Discrete Distributions
Continuous Univariate Distributions—1
Continuous Univariate Distributions—2
Continuous Multivariate Distributions
JUDGE, HILL, GRIFFITHS, LÜTKEPOHL and LEE • Introduction to the Theory and Practice of Econometrics, *Second Edition*
JUDGE, GRIFFITHS, HILL, LÜTKEPOHL and LEE • The Theory and Practice of Econometrics, *Second Edition*
KALBFLEISCH and PRENTICE • The Statistical Analysis of Failure Time Data
KISH • Statistical Design for Research
KISH • Survey Sampling
KUH, NEESE, and HOLLINGER • Structural Sensitivity in Econometric Models
KEENEY and RAIFFA • Decisions with Multiple Objectives
LAWLESS • Statistical Models and Methods for Lifetime Data
LEAMER • Specification Searches: Ad Hoc Inference with Nonexperimental Data
LEBART, MORINEAU, and WARWICK • Multivariate Descriptive Statistical Analysis: Correspondence Analysis and Related Techniques for Large Matrices
LINHART and ZUCCHINI • Model Selection
LITTLE and RUBIN • Statistical Analysis with Missing Data
McNEIL • Interactive Data Analysis
MAGNUS and NEUDECKER • Matrix Differential Calculus with Applications in Statistics and Econometrics
MAINDONALD • Statistical Computation
MALLOWS • Design, Data, and Analysis by Some Friends of Cuthbert Daniel
MANN, SCHAFER and SINGPURWALLA • Methods for Statistical Analysis of Reliability and Life Data
MARTZ and WALLER • Bayesian Reliability Analysis
MASON, GUNST, and HESS • Statistical Design and Analysis of Experiments with Applications to Engineering and Science
MIKÉ and STANLEY • Statistics in Medical Research: Methods and Issues with Applications in Cancer Research
MILLER • Beyond ANOVA, Basics of Applied Statistics
MILLER • Survival Analysis
MILLER, EFRON, BROWN, and MOSES • Biostatistics Casebook
MONTGOMERY and PECK • Introduction to Linear Regression Analysis
NELSON • Applied Life Data Analysis
OSBORNE • Finite Algorithms in Optimization and Data Analysis
OTNES and ENOCHSON • Applied Time Series Analysis: Volume I, Basic Techniques
OTNES and ENOCHSON • Digital Time Series Analysis